Mars: The Next Step

Mars from Viking 2 (NASA/JPL).

Mars: The Next Step

Arthur E Smith

Adam Hilger
Bristol and New York

British Library Cataloguing in Publication Data

Smith, A. E. (Arthur Edward), *1920–*
  Mars: the next step.
  1. Mars. Exploration
  I. Title
  523.4'3

  ISBN 0-85274-026-3

Library of Congress Cataloging-in-Publication Data

Smith, Arthur (Arthur E.)
  Mars: the next step/Arthur E. Smith.
  148 p. 216 cm.
  Bibliography: 4 p.
  Includes index.
  ISBN 0-85274-026-3
  1. Mars (Planet)–Exploration. 2. Astronautics in astronomy.
I. Title.
QB641.S65 1989
523.4'3–dc19   88-35092
                                                                                CIP
Consultant Editor: **Professor A E Roy**, University of Glasgow

Published under the Adam Hilger imprint by IOP Publishing Ltd
Techno House, Redcliffe Way, Bristol BS1 6NX, England
355 East 45th Street, New York, NY 10017-3483, USA

Typeset by KEYTEC, Bridport, Dorset
Printed in Great Britain at The Bath Press, Avon

Contents

Foreword by Heather Couper   vii

Introduction   ix

1   The Red Planet   1

2   The First Probes   9

3   The Search for Life   34

4   The New Generation   45

5   The First Explorers   61

6   The Big Boosters   86

7   The Human Factor   103

8   The Cost   113

9   The Transformation   121

Bibliography   138

Index   143

Foreword
by Heather Couper

On 30 October 1938, listeners to CBS radio in New York were aghast to hear that an alien invasion had apparently begun. The terrified studio announcer reported that Martians had landed in New Jersey, and were now converging on New York, devastating everything that lay in their path. On hearing the news, people took to the hills, the roads—anywhere. The roads out of New York were jammed solid. But not everyone could get away. And those who stayed at home were horrified to hear the announcer's final screams as the Martians burst into the studio . . .

It was a hoax, of course: a clever adaptation of H G Wells' classic story *The War of the Worlds* by the 23-year-old Orson Welles. It guaranteed him instant fame—and a glittering future as a director in films and the theatre.

1988 was the 50th anniversary of the scare, and the event was celebrated by repeats of the broadcast worldwide. Astonishingly—even in these supposedly enlightened times—scenes of panic and pandemonium broke out all over again. Clearly, the belief in life on Mars is firmly embedded in our psyche.

As Arthur Smith describes in this fascinating book, our preoccupation with life on the Red Planet began in 1877, when Giovanni Schiaparelli reported that Mars appeared to crisscrossed by a network of 'channels'. Although most astronomers never went as far as Percival Lowell—who was convinced that these were canals constructed by a dying race of intelligent Martians—there was always a lingering hope that Mars might harbour some form of life. I well remember that several of the first astronomy books I read (written in the late 1950s) were quite confident that we would find primitive lichens or mosses on the martian surface.

The cold realism of the space era certainly dented this optimism. The first good space-probe images of Mars, sent back in 1965 by Mariner 4, showed a crater-scarred world more reminiscent of the dead Moon than of our living Earth. But when Mariner 9 went into orbit about Mars—and the global dust storm raging at the time finally cleared—hopes picked up again. Here were parts of the planet that had never been seen before: gigantic volcanoes, gargantuan canyons, and—most important of all—narrow, winding gullys that appeared to have been cut by running water. If there had ever been water on Mars, then life might just have got started. The way was clear for the Viking mission, with its sophisticated life-detection experiments. When the Viking landers *didn't* find any trace of life, or even of the organic compounds that form the precursors of life, scientists were bitterly disappointed. The negative findings coloured their whole attitude to finding life elsewhere in the Universe. If not on Mars—which isn't really such a bad place—then where?

More than a decade has passed since the Viking results, and scientists have had time to come to terms with the findings. There is now a new optimism that one day, we *will* find life on Mars. This time, there can be no doubt—because it will be human life.

Arthur Smith describes in detail the new initiatives that are being taken to return to Mars. Both the Soviet Union and the United States have far-reaching programmes that will culminate—hopefully—in a manned landing early next century. Already, teams on Earth are preparing for living on Mars. In Arizona, a simulated Martian environment, Biosphere II, has been carefully built up over a number of years, and is shortly to admit its first human residents. One day, we will have the real thing: permanent human colonies on Mars.

Why? Because—as Arthur Smith relates in the later chapters of this book—it is inevitable: as inevitable as our migrations to every corner of planet Earth. To our descendants, Mars will be a real 'new world' to explore. It will give the human race its first experience of living on a truly alien world, a world that is totally independent of Earth.

And in more ways than one, Mars is our 'next step'. Just before his death, the great German space pioneer Krafft Ehricke said: 'If God wanted man to become a spacefaring species, he would have given man a moon'. To which the anthropologist Ben Finney adds: 'If God wanted humans to migrate to other stars, he would have given us a moon, Mars, asteroids, the outer planets and their varied satellites, and the Oort Cloud of comets'. Mars is the next step on our way to the stars.

Introduction

Despite a virtual generation of studies based on the newly devised technology of space, Mars remains something of an enigma to man. Fly-by, orbiting and landing missions by increasingly sophisticated space probes have revealed more about the Red Planet in the last thirty years than had been gleaned in all previous centuries, but there are still many mysteries to be solved.

They range from the question of where Phobos and Deimos, the diminutive martian satellites, came from to the whereabouts of the water which once sculptured much of the planet's surface. These and many more questions will only be answered after years, perhaps decades, of effort; and because the planet and its satellites preserve much of the ancient history of the Solar System, they are important questions for science. Indeed, the scientific curiosity of man and the rivalry between East and West guarantee that during the next two or three decades Mars will occupy the central niche in space research that the Moon occupied in the 1960s.

Both the United States and the Soviet Union have given notice of their intention to mount manned expeditions to Mars. Neither has yet demonstrated any hardware that would make such an accomplishment possible—apart from the large Soviet booster rocket, the Energiya—but the intentions of both nations are serious and I fully expect some sort of manned expedition to set out early in the next century.

By then, the Soviet cosmonauts manning the Mir space station and its successors will have acquired far more experience of prolonged weightlessness than their American rivals, whose space station will not be in orbit before the middle of the 1990s. That experience will stand them in good stead in the biomedical sense, but so powerful is the American economy and so far advanced its technology that the US hardware will almost certainly be superior. The problems that the US technical establishment has experienced with the shuttle since the Challenger disaster represent no more than an historical aberration and not a sign of weakness.

But it is not the main purpose of this book to discuss the respective merits of the US and Soviet approaches to manned exploration of Mars. Indeed, it would be far more sensible and satisfactory if the two efforts could be combined, with a wider global participation even more desirable. The primary aim is to demonstrate that the only feasible colonization of another planet in our Solar System will be on Mars.

The conquest of the Moon was merely a precursor to a more dramatic step forward, which will take men and women to an outpost which might eventually become a second home for the human race. Mars is not an ideal habitat—the science fiction writers who envisaged an environment in which men could live with ease could hardly have got it more wrong. But with the aid of technologies that have only just begun to be understood, it may be possible in the next century or two to produce an environment on Mars that will support life.

The unfortunate side-effects of industrialized society are giving cause for concern on Earth, but it may be that advanced technology can be used on another planet to make life flourish where it does not currently exist. Such technology could also be highly beneficial to the ecology of the Earth.

It is more than a coincidence that the environmentalist movement which sprang up in the late 1960s followed closely on one event—the Apollo 8 circumlunar mission in December 1968. For the vast majority of the Earth's population the word 'ecology' was meaningless until astronaut Frank Borman turned round to the planet he had just left and described it as 'spaceship Earth'. From that day on, the movement to deal more justly and more carefully with the resources of the Earth's closed system gained support rapidly. It was realized as never before that the resources of the Earth are finite and, while some are renewable as long as they are carefully managed, many were being dangerously abused by man's growing numbers. This is one of the strangest and perhaps most valuable spin-offs from space exploration—the opportunity to look back on Earth and see it as it really is, a tiny oasis of life in an otherwise lifeless Solar System.

It also enables the human race to gain a true perspective on the rest of the Solar System. For the first time in the million or so years that the race has been evolving, we can look outwards and see our planetary neighbours in terms of resources. Exploration has shown that certainly the Moon and Mars and perhaps other bodies contain readily available elements such as oxygen, aluminium and iron. Compounds such as carbon dioxide, carbon monoxide, methane, water, hydrogen peroxide and oxides of nitrogen are also either available or can be manufactured from local materials without too much difficulty once a base is established. These benefits are there for the taking by any generation willing to take the risks involved.

Critics of a manned Mars expedition would suggest that the huge sums of money involved—undoubtedly running into hundreds of billions of dollars—would be better spent on terrestrial research projects. At a time of stringency in national research budgets, this argument may appear to carry some weight—until one examines the evidence. It is clear that space studies have given rise to huge advances in our understanding of the physical environment of the Earth, and will continue to do so as our sphere of research expands further into the interplanetary domain. The aims of the terrestrial and space communities are in no way conflicting, and the potential for mutual gain should never be underestimated.

The development of Mars as a second home for mankind will be slow, but in the long term I believe that it will provide a new goal for the human race. It will also be the source of new scientific advances that are currently undreamed of, comparable to those that have transformed our knowledge of the Universe and its physical laws in the twentieth century.

Acknowledgments

I would like to thank members of the Public Affairs staff of NASA at Headquarters and the Jet Propulsion Laboratory, Marshall Space Flight Center and Ames Research Center for valuable assistance, research material and illustrations. My thanks also go to Carter Emmart and D R Woods for permission to use their drawings.

1 The Red Planet

Mars alone enables us to penetrate the secrets of astronomy which otherwise would remain forever hidden from us.
Johannes Kepler

Mars is the fourth planet from the Sun. Its orbit lies between those of the Earth and Jupiter, but this was unknown to the ancients who merely recognized it as one of the 'wanderers' (in Greek, *planetes*). Five planets were known in antiquity: Mercury, Venus, Mars, Jupiter and Saturn; Uranus, Neptune and Pluto were unknown until the era of the telescope. They were called planets because the ancient Greeks, and perhaps men thousands of years earlier, noticed that these relatively bright objects do not stay in fixed positions in the sky as the stars do but wander among the constellations roughly in the plane of the ecliptic—the plane in which the Earth orbits the Sun.

During millennia in which there were no bright city lights to obscure the night sky, hunters or shepherds must have noticed these bright lights and the shifting relationship they had with the fixed stars; indeed, several cultures included them in their pantheons of gods.

Mars had a special place in this nightly display, for unlike the other planets it has a reddish tinge which led the astronomer–astrologer priests of ancient Babylon 3000 years ago to name it Nergal, after their god of death and pestilence.

For the same reason, the ancient Greeks came to identify the planet with Ares, their god of war, presumably because of the association war has with blood and fire. The Romans, who took their gods from the Greeks, similarly called the planet Mars, after their god of battle.

The red glare of Mars is particularly noticeable at the times of opposition, when the Earth is directly between Mars and the Sun. When an opposition is 'favourable'—that is to say, when the distance between the two planets is a minimum—Mars can be one of the most striking objects in the sky, as it was in the autumn of 1988. Its magnitude then as measured by astronomers on their logarithmic scale was −2.8, which is exceeded in the night sky only by the Moon and

Venus. Venus can reach a magnitude of −4.4, which is four times as bright as −2.8.

But even at these favourable oppositions, the apparent diameter of the planet is only 25.1 seconds of arc. By contrast, the Moon's maximum diameter as seen from Earth is 33′ 36″, or eighty times as great. As a result, the ancients were totally unable to observe any detail on the face of Mars. Even the most eagle-eyed observer sees it as only a tiny reddish blob of light in the sky with no distinguishable markings.

However, Mars played an absolutely vital role in the development of astronomy even before telescopes were invented. The two key players in this development were Tycho Brahe, the Dane who was the supreme naked eye observer, and Johannes Kepler, the most important theorist of the early 17th century. Naked eye observations, in which the changing position of Mars was measured with great accuracy, revealed a disturbing fact.

Mediaeval thinking inspired by Aristotle maintained that heavenly bodies all existed in a state of perfection, and therefore their orbits must be circular, since a circle was 'perfect'. But the observations made by Tycho with unprecedented precision over a period of 35 years during the latter half of the 16th century, using instruments he built himself, could not be reconciled with a circular orbit.

Under the Ptolemaic system, dating from the second century AD, in which the Sun, the Moon and the planets had been assumed to orbit round a stationary Earth, it had long been realized that simple circular orbits did not fit the observed motions of the heavenly bodies. 'Epicycles' were added to the motions of the planets, i.e. they were said to be describing smaller circles in space as they revolved in their circular orbits around the Earth. As observations improved it became necessary to increase the number of epicycles until in the end it was assumed that 39 'wheels' were involved in the motions of the Solar System. A fortieth was needed to explain the motions of the fixed stars.

These attempts to 'save the phenomena' became unrealistic when Copernicus replaced the geocentric universe with a heliocentric Solar System.

To be absolutely accurate, Copernicus believed that the Earth and the other planets revolved around a point removed some distance from the Sun. He was still obsessed with 'uniform circular motion' and his system, rather than doing away with the idea of epicycles, increased their number to 48. This was necessary because he still believed that all

2

orbits were circular and this did not fit the observations.

In an intellectual struggle which lasted many years, Kepler was able to fit Tycho's observations not to circular orbits but to elliptical ones. This shattered mediaeval notions of astronomy and laid the foundations in the next few decades for the great discoveries of Sir Isaac Newton.

Without actually citing gravity, Kepler suggested that a force emanating from the Sun must be responsible for the motion of the planets. In doing so he completed the transition from a geocentric universe to a heliocentric universe which Copernicus had started. In the end, he realized that the only way in which the observations could be made to fit was by assuming that the orbit of Mars was an ellipse, with the Sun at one of the foci. From this realization sprang the whole of Kepler's insight into the motions of heavenly bodies, published in his seminal work, of 1609, *Astronomia Nova*, and eventually enshrined in his three laws of planetary motion.

Not least of Kepler's achievements was the fact that he was the first to contend that Mars and the other planets were material bodies and therefore imperfect like the Earth. Before *Astronomia Nova* it was universally assumed that change could occur only in the 'sub-lunary sphere' and that everything in the heavens was immutable and perfect.

It is difficult to overestimate the part that Mars played in this complete overturning of the mediaeval concept of the Universe, although in many ways Kepler, with a profound belief in occultism and the mystical attributes of the heavenly bodies, remained a mediaeval man himself. Even so, he was able to comment: 'Mars alone enables us to penetrate the secrets of astronomy which would otherwise remain hidden from us.'

It is ironic that this central figure in the history of astronomy actually knew very little about the true physical nature of Mars when he died in 1630.

Telescopes had been invented two decades earlier and Kepler undoubtedly made telescopic observations of Mars, but the resolving power of those early instruments was not sufficient to show any details. In truth, surface details were not really understood until the arrival of the martian space probes of the 1970s.

Galileo, after his pioneering observations of Jupiter and the Moon, also turned his telescope on Mars. However, with his crude instruments—a cheap modern pair of mass-produced binoculars would be far superior—he was only able to note that Mars did not appear round. This was because, being some way from opposition, it was

3

gibbous and only part of the disc was illuminated as seen from Earth. The first drawings of the disc were made by the Italian, Francesco Fontana, in 1636, but the markings he portrayed do not resemble anything seen later.

About 50 years after Galileo's first observations, the Dutchman, Christian Huygens, produced the first detailed drawings of the surface of Mars from observations he made on 28 November 1659. As crude as the drawings were, they showed detail that is recognizable when compared with much later maps made from telescopic observations as much as 300 years later. In particular, he showed the dark triangular area which later came to be known as Syrtis Major.

Huygens was also able to report that the planet rotated, although its period of rotation was not measured until Giovanni Domenico Cassini estimated it at 24 h 40 min in 1666 after observing the rate at which markings moved across the disc. This was remarkably accurate, for it differs by only three minutes from the true value.

Cassini noted the presence of polar caps in his observations, and the British astronomer and discoverer of Uranus, Sir William Herschel, was the first to postulate that they were of water ice, like the terrestrial polar caps. It was Herschel who first noted visual evidence that Mars had an atmosphere and he also measured the tilt of the polar axis.

It was, of course, during the 17th century that Sir Isaac Newton made his revolutionary contribution to the theories of physics, with new knowledge of the planetary orbits acquired by Kepler in his study of Mars playing a central role in this upheaval. Armed with the Keplerian laws, Newton 'stood on the shoulders of giants,' as he himself said, and was able to reach his conclusions on the nature of gravity in a manner inconceivable without those advances.

As telescopes improved over the centuries, so the appearance of the surface of Mars came to be known more and more fully.

Phenomena that were observed included yellow, blue, white and grey clouds, waxing and waning of the polar caps, seasonal changes in the colour and extent of the dark areas, a green haze and bright transient spots.

The most significant phenomenon to be observed was the so-called 'wave of darkening' that sweeps across the planet in phase with the melting of the polar caps. This phenomenon, first discussed by the American astronomer Percival Lowell in 1894, moves away from the melting polar cap in the summer hemisphere at a rate of 35 km a day. As recently as the 1950s it was possible for this wave of darkening to be discussed as possible proof of vegetation burgeoning under the influence of water being released from the polar caps.

Only with the successful flight of Mariner 9 in 1971 was it shown that the wave of darkening is probably no more than an optical illusion.

Astronomers during the 19th century followed a convention that had first been established in lunar observations by giving names such as continent, sea, ocean, island and strait to markings seen on the surface of Mars. Telescopic observations of a faint object are notoriously difficult to analyse, for each observer puts a different interpretation on the details that he sees. Some see clear-cut divisions between different areas and interpret them as coastlines, whereas others see margins in a more indistinct manner.

Whatever the reason, the 19th century observations helped to build up the myth of a fairly benign environment, where martian life might flourish. The fact that dark regions had indistinct edges was interpreted to mean that they were shallow seas with ill defined shores, like the large tidal swamps found bordering tropical seas on Earth. Changes in the extent of the markings could then be seen as small oscillations in the level of the oceans.

Later, astronomical observations with more sophisticated instruments revealed a little more about the nature of the martian surface. The amount of sunlight reflected by the reddish areas, for instance, was 15 to 20 per cent, which made them similar in appearance to terrestrial deserts.

Lowell was responsible for perhaps the longest and most confusing debate in the history of astronomy since the Middle Ages—the dispute about the 'canals' of Mars. In 1877 the Italian astronomer Giovanni Schiaparelli described *canali* which he had observed on the martian surface. The Italian word means 'channels', but perhaps inevitably it was translated into English as canals.

The canals were popularized in 1900 by Lowell, who saw them as unerringly straight lines which could not be of geological origin and must be artefacts constructed by intelligent Martians.

At this time, of course, there was very little scientific evidence on the nature of the martian atmosphere and climate, so the idea of an advanced, intelligent race on Mars could seem just as likely as the idea that no such race existed. Lowell suggested that the lines, which were much too wide to be mere watercourses, were bands of vegetation bordering canals bringing water from the polar caps to the dry equatorial regions of a desiccated planet.

The story of the martian canals is well enough known not to need

5

extensive description. It is sufficient to say that the canals simply do not exist. Laboratory experiments have shown that in seeing conditions similar to those experienced during telescopic observations of Mars, faint markings tend to appear as lines to the human eye.

The 'canals' of Mars, then, were an optical illusion, but during the half century or so in which the debate raged no-one could absolutely deny their existence. With the aid of numerous science fiction stories, such as *The War of the Worlds* by H G Wells, published in 1898, they even generated a popular belief in the existence of Martians who might be a menace to the Earth.

The popularity of this type of fiction, with the books of Edgar Rice Burroughs, for one, selling in their millions, more or less established in popular mythology the acceptance that intelligent beings would one day be found on Mars—or would perhaps invade the Earth.

This myth had been given added force when a planet-wide dust storm was observed on Mars in 1904. The storm produced a shape like the letter W—identified only 70 years later as the tips of the great volcanoes—on the surface of Mars, and this provoked a lively debate about the possibility that the Martians were preparing for war.

The well known episode in the 1930s when the actor Orson Welles broadcast an adaptation of the H G Wells story and caused a panic in many American states is proof that the idea of warlike Martians posing a threat to the Earth was strongly established even in modern times, difficult as it is for the present generation to understand.

However, by the middle of the 20th century, when space exploration began, the basic facts about Mars were known to science, even though the existence of life had not been proved or disproved.

Mars is not a large planet. With a radius at the Equator of 3398 km, slightly more than half that of the Earth, it has a mass of 6.421×10^{23} kg, which is not much more than a tenth of the terrestrial mass.

Like the Earth, Mars has a polar axis which is tilted in relation to the plane of its orbit. In fact, the angle of inclination in each case is remarkably similar—23.45° for the Earth and 23.98° for Mars—but this is not believed to be anything other than coincidence. In each case the inclination varies widely over a period of thousands of years, and it just happens that in this era the inclinations are similar.

A result of the inclination of the axis is that Mars has four seasons, just like the Earth, though they are different in length because the martian year is longer. Mars is at aphelion (furthest from the Sun) during the southern winter and northern summer, a period which lasts

182 Earth days. It is at perihelion during southern summer and northern winter, a period of 160 days. Southern spring and northern autumn last for 146 days, and southern autumn and northern spring for 199 days. This means that the southern hemisphere receives greater insolation in its summer and is subjected to winter cold for a longer time. As a result, the southern polar cap shows greater variations in size from its maximum extent of 30° north than the northern cap.

Another coincidence is that the period of rotation and therefore the length of the day on Mars are almost exactly the same as on Earth.

The Earth's sidereal period (the time in which it returns to the same position relative to the stars) is 23.9345 h, while Mars rotates in 24.6229 h. There is no synchronicity between these day lengths, so the explorers of Mars, to avoid confusion, will have to operate by a martian calendar, with the day being called a 'sol'.

As Mars is further from the Sun, its orbital period and therefore its year is longer than ours. It orbits the Sun in 687 days in an elliptical orbit which is more eccentric than any other planet's except those of Mercury and Pluto.

While its mean distance from the Sun is 277 900 000 km—about 50 per cent greater than the Earth's—the semi-major axis of the orbit is 249 100 000 km and the semi-minor axis is 206 700 000 km. As will be seen later, this eccentricity of 0.0934, leading to a variation of almost 10 per cent in the distance from the Sun, has profound implications for all missions to Mars, both manned and unmanned.

Eventually, a calendar for Mars will have to be devised. The most recent attempt among many to reconcile the different day and year lengths has been made by Thomas Gangale (1988 *Spaceflight* **30** July 278–83). In what he calls the Darian calendar there are 24 months, each of either 27 or 28 martian days. Just as on Earth there are four weeks in the month but there are 96 weeks in the year.

When humans land on Mars they will need some form of timepiece as well as a calendar. It might be psychologically distracting to be linked for fifteen months or so to a terrestrial timescale gradually getting more and more out of phase with sunrise and sunset. In these days of digital timekeeping it will be a simple matter to re-program the ship's computers and even wrist watches to run slightly slower so that they register 24 'hours' in 24 h 37 min 22.56 sec.

Radar measurements taken from Earth in recent decades have established that the planet is not a perfect sphere. Its polar radius at 3380 km is 18 km less than its equatorial radius, so the polar flattening is $\frac{1}{192}$, considerably more than the Earth's $\frac{1}{300}$. This polar flattening is brought about by the rotation of the planet, which causes an equatorial bulge.

But with all these known facts, until it became possible to send packages of observational instruments to Mars, the true nature of the planet's surface and atmosphere was quite unknown. The construction of large rockets and miniaturized instruments opened up a completely new chapter in the study of Mars and scientists both in the United States and in the Soviet Union were quick to take up the challenge. The first tentative missions to Mars began only three years after Sputnik 1 and the first success only four years later.

2 The First Probes

The exciting thing about comparative planetology is that it will permit us to unfold the lost part of the Earth's history, now largely obliterated by erosion, mountain building, and other processes.
S E Dwornik

By the early 1960s Earth-based telescopic observations had established at least some of the basic parameters of the martian environment. The orbits of the planet itself and its two tiny satellites were known with fair precision and reasonable assumptions had been made about the composition of the planet and its atmosphere.

But this was only scratching the surface and those scientists interested in planetology knew that they had to take their instruments closer if the secrets of Mars were to be unravelled. So began a programme of exploration by more and more sophisticated robot spacecraft which is continuing and will continue at least until a manned expedition is mounted.

The first attempts to reach Mars were made by the Soviet Union as early as 1960, when two probes were launched. On 10 and 14 October two rockets were launched from the Tyuratam 'cosmodrome' in the Central Asian republic of Kazakhstan, but in each case the spacecraft failed even to reach orbit around the Earth.

These two failures were only revealed by the United States much later because Moscow had alleged that they were secret American military launches.

In theory, if you have a powerful enough rocket it is possible to launch a spacecraft towards Mars at any time. But in practice, while using the current and foreseeable generations of chemical rockets, the relationship between Earth and Mars places restraints on interplanetary journeys. The two planets must be in positions such that when a rocket is launched from Earth it goes into a transfer orbit round the Sun which intersects the orbit of Mars. The heliocentric orbit is elliptical, with its aphelion, or most distant point from the Sun, coinciding with or crossing the orbit of Mars.

The intersection must take place at a time when Mars is occupying that particular point on its orbit. The planets are in positions where this is possible approximately every twenty six months, when Mars is at or near opposition. Opposition occurs when a straight line drawn from the Sun to the Earth would also pass through or at least near Mars. The few weeks in which it is possible to launch probes to Mars are known as the launch window.

Outside the launch window either an impossibly large rocket or a zero payload would be required for an attempt. Even during launch windows the energy requirements vary widely from one year to another, depending both on the weight of the payload to be launched and on the distance between the two planets at opposition.

Between 1964 and 1976 a total of 13 operative payloads reached Mars; the transit time has varied from 131 days for Mariner 7 in the favourable 1969 window to 333 days for Viking 2 in 1975 and 1976. The Russian missions have varied from 188 to 249 days.

The distance between Earth and Mars at opposition varies in a fairly regular cycle of about 15 years because the Earth's orbit is almost circular but Mars orbits the Sun in an ellipse and it is much closer to the Earth when its perihelion coincides with opposition. The cycle is not completely regular because the orbit of Mars is inclined to the plane of the ecliptic by 1.85° and closest approach does not always coincide exactly with opposition. For instance, the distance was at a minimum of about 56 000 000 km in the autumn of 1988. Mars will not be as close again until the year 2003.

The launch windows in the middle 1960s were all unfavourable, but at the launch window after the 1960 failure the Soviet Union made no fewer than three attempts. Once again, two payloads failed to achieve Earth orbit, on 24 October and 4 November 1962.

Between these two failures was the partial success of Mars 1, a development of the Venera craft with which the Soviet Union was then studying the planet Venus. It was a relatively small spacecraft by Soviet standards—a mass of just under 900 kg was the largest that could be placed on a trajectory to Mars by the Soviet launcher of the time. Mars 1 was intended merely to fly close to Mars, taking a few hundred pictures as it did so and perhaps measuring composition and temperature of the atmosphere. It would then continue on its elliptical orbit around the Sun, but unfortunately the radio, the craft's only link with Earth, failed on the outward journey and no data or pictures were received. Mars 1 did fly within 193 000 km of its planetary namesake, so it could be accounted a partial success at least from the guidance point of view.

With their next attempt at the next launch window in 1964 the Soviet team were more cautious and called the craft Zond 2 rather than Mars 2. The word Zond means 'probe' and was applied to craft launched towards the Moon and Venus as well. Moscow did announce that Zond 2 was to study Mars when it was launched on 30 November 1964, but it too suffered from a faulty radio and although it passed within a creditable 1500 km of Mars it transmitted no information about the planet.

Obviously something had to be done to put right the faults that were plaguing the Mars programme. The solution was to launch a back-up of Zond 2 on an engineering test flight.

Zond 3 was launched on 18 July 1965, outside a launch window, but it was still transmitting when it crossed the orbit of Mars so some progress was made. Mars, of course, was tens of millions of miles away when the orbit was crossed, but on its outward journey Zond 3 was able to test its camera systems by taking fairly high quality pictures of the Moon.

The decision was then taken to go to a new and heavier spacecraft, based on the greater launch capacity of the big Proton rocket. This 1000 tonne rocket can place 20 tonnes into Earth orbit and more than $4\frac{1}{2}$ tonnes on an interplanetary course, and it was first used to aim for Mars after two launch windows had been missed.

Once again three spacecraft bound for Mars left Tyuratam's launch pads between 10 and 28 May 1971.

The first failed to leave Earth orbit, was named Cosmos 419 as a cover for its failure, and re-entered the Earth's atmosphere after two days. But both Mars 2 and Mars 3 were successfully placed on trajectories that brought them to Mars after journeys of just over six months.

Each of the craft had an orbiter with a mass of about 2 tonnes after fuel was burned to place it into orbit, and a lander of 635 kg. Mars 2 went into orbit successfully, the lander was detached and its retrorocket fired and it headed towards a landing point in the southern hemisphere. Unfortunately, no signals were received from the lander, which was presumably destroyed on impact.

Mars 3, which was a twin of Mars 2, had better luck. It was injected into orbit round Mars on 2 December 1971, and later the same day its lander made a soft landing in another part of the southern hemisphere. The lander did survive the landing but not for long; the orbiter, which was acting as a relay for the weak signals from the surface, broadcast data for a mere twenty seconds. The signals from the lander may have been cut short by the effects of a dust storm that was raging at the time.

However, the orbiters at least managed to elucidate some of the planet's secrets, being equipped with a full range of instruments as well as two cameras with wide angle and telephoto lenses. From their 48.9° orbits round Mars they continued to operate for about four months, sending back images of the surface in digital form.

The photographic system was conventional: an ordinary photographic emulsion system was exposed and the resulting pictures were developed automatically. An electronic system then scanned the images, breaking them down into 1000 by 1000 picture elements, before beaming back to Earth the million data points for each picture. This did not give a great deal of information, for the surface of Mars was obscured at the time by a dust storm.

Nevertheless, it was a start and some scientific data about the composition of the martian rocks were obtained by the instruments on the orbiters.

Atomic hydrogen and oxygen were identified in the upper atmosphere, surface relief was studied by measuring the amount of carbon dioxide along a sighting line, and water vapour concentrations were mapped. Temperatures in the atmosphere were found to vary between -110 and $13\,°C$ and differences of relief as great as $4\,km$ were observed.

In the next launch window, in the summer of 1973, energy requirements were higher, and it was not possible to launch both an orbiter and a lander on one Proton rocket. As a result, the 1973 fleet consisted of no fewer than four separate spacecraft. The orbiters, Mars 4 and 5, went first, on 21 and 25 July, and they were followed by the landers, Mars 6 and 7, on 5 and 9 August. Unfortunately, almost everything that can go wrong on a space mission did go wrong in February and March of the following year when the flotilla arrived.

First to arrive was Mars 4, but its retro-rocket, designed to reduce its speed so that it would be captured into an orbit round Mars, failed to work. It continued in its heliocentric orbit after taking and transmitting to Earth some pictures from as little as 2200 km from the planetary surface.

Two days later Mars 5 went into a martian orbit with an inclination of 35 degrees to the equator, a periapsis of 1760 km and an apoapsis of 32 500 km. Its two television cameras sent back better images of the surface and its instruments detected levels of water vapour four to eight times greater than those detected by Mars 3.

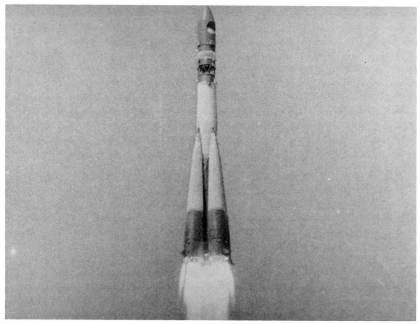

The Soviet Union surprised the West with this launch vehicle in the 1957 launch of Sputnik 1. It is still in weekly use in military and civilian space programmes, but the Energiya rocket, many times larger, will be needed for Mars missions (Novosti).

Mars 6 and 7 both failed to produce any results, the first because its lander ceased to transmit just $2\frac{1}{2}$ minutes before touchdown and the latter because the lander had an on-board failure and missed the planet altogether.

Images from the Mars 5 cameras were used by both Soviet and American planetologists, together with US pictures, to draw maps of some martian features. This 1974 campaign signalled the end of the Soviet Union's interest in Mars for some time and it did not resume until the Phobos launches of 1988.

Meanwhile the American campaign to explore Mars has been both more intensive and less expansive than the Soviet campaign. Fewer spacecraft were much more intensively instrumented and produced much more in the way of results.

13

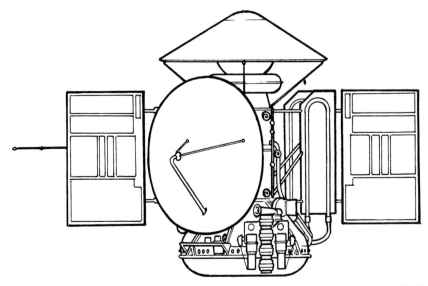

Mars 6. Even in the early seventies unmanned spacecraft were remarkably complex, but the new Soviet probes in the Phobos programme are at least an order of magnitude more complex (Novosti).

They began with the Mariner Mars mission in 1964. This was intended to be a twin flight of payloads of the powerful Atlas–Centaur rocket, each of several hundred kilogrammes, but as in other programmes the delays in the development of the Centaur upper stage meant that a much smaller craft had to be substituted.

The substituted Atlas–Agena could inject 261 kg into the martian trajectory in the favourable 1964 geometry and so this was the weight of Mariners 3 and 4. They were based on the Ranger spacecraft which had succeeded in lunar research, but they were much more complex, with over 130 000 parts in each spacecraft.

One addition was a television camera, since a primary aim of the two missions was to photograph the martian surface, seen up to then only by means of telescopes and through the murky Earth atmosphere. It was not an advanced camera by later standards; it was planned to take only 21 pictures and store them in digitized form, to be transmitted later at the extremely slow rate of one picture every 8 h 20 min. However, it was the first attempt to photograph the surface of Mars and the first to succeed.

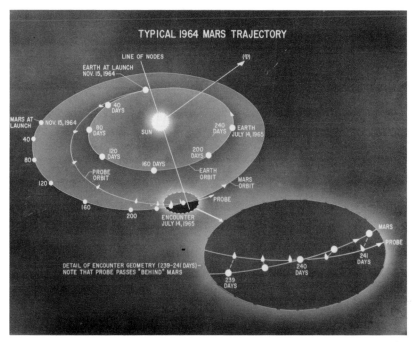

A diagram illustrating a 1964 Mars trajectory. The Mariner 4 flight was only slightly different. The craft left the Earth on 28 November 1964 and encountered Mars on 15 July 1965. It is still in orbit round the Sun (NASA).

Mariner 3 was launched on 5 November 1964 (four days after Mars 1) and at first seemed to be going well. The Atlas rocket burned normally, but the shroud that covered the payload failed to separate. Although the Agena fired and injected the Mariner into a course towards Mars the mission failed because the spacecraft could not work with its solar panels and antennas covered by the shroud.

A quick re-design of the shroud enabled a more successful mission to begin only 23 days later, when Mariner 4 began its journey to the Red Planet.

This time there was no mistake and although the Atlas–Agena placed Mariner 4 into a trajectory which would have missed by 240 000 km, it was possible to correct the course by a brief burn of the hydrazine-powered engine on 5 December.

15

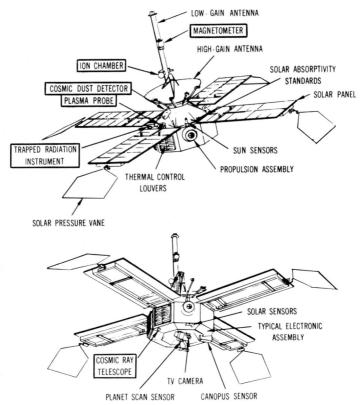

Mariner 4 was a simple spacecraft by later standards and had only a few minutes to take close-up pictures of Mars. Yet it revolutionized planetologists' ideas of the planet, for it revealed a cratered surface which was more Moon-like than Earth-like and which must have been very ancient (NASA).

Closest approach was on 15 July 1965, beating Mars 1 by four days, with an acceptable miss distance of 9851 km after a voyage of 230 days. The 30 cm focal length Cassegrain telescope of the television system had been designed to cope with a range of illumination of 30 to 1, ranging from full solar illumination to near total darkness at the terminator. It duly took 22 pictures of the martian surface at ranges which varied from 17 000 to 12 000 km. They were history-making images which revealed that the enigmatic surface of the planet was

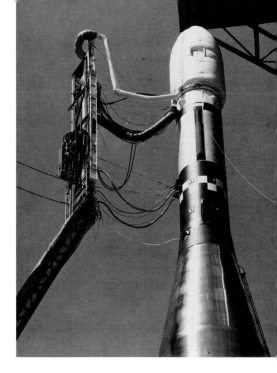

The Mariner 4 spacecraft on top of its Atlas–Agena launch vehicle at Cape Canaveral. The insulating blanket, designed to keep the spacecraft at an even temperature, was removed before launch (NASA/KSC).

liberally sprinkled with craters, at least on the 1 per cent of the surface that the Mariner covered.

There were 70 clearly distinguishable craters ranging in diameter from 4 to 120 km, with rims rising about 100 m above the surrounding surface and with depths of many hundreds of metres. It was a surface remarkably similar in the size distribution of the craters to the uplands of the Moon.

Mariner 4 was also able to put an upper limit on the martian magnetic field of not more than one thousandth of the terrestrial field. There was therefore no surprise in the discovery that there is no radiation belt around Mars because a much stronger magnetic field is needed to trap the energetic particles which make up a belt such as surrounds the Earth. But Mariner 4 did find that interplanetary dust is more abundant in the vicinity of Mars than it is near the Earth.

It also established that the density of the atmosphere was only about 1 per cent that of the Earth's. With a data transmission rate of only 8.33 bits per second, Mariner 4 was still able to return to Earth a healthy package of data and images that radically transformed the conventional view of Mars as somehow comparable with the Earth.

America, like the Soviet Union, missed the next launch window but

17

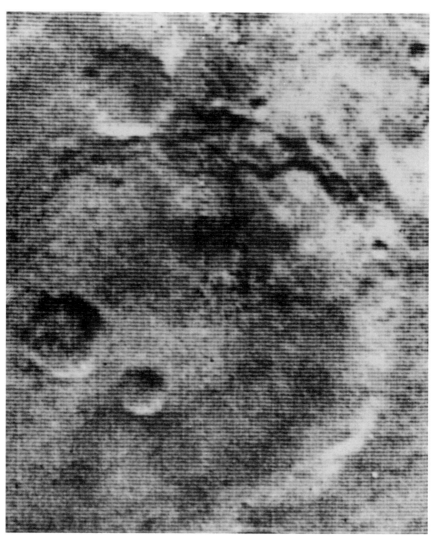

The poor resolution and coarse reconstruction of this Mariner 4 picture of the martian surface demonstrate just how far the Mariner 9 and Viking missions progressed over the pioneering effort. Yet Mariner 4 was a remarkable state-of-the-art achievement (NASA/JPL).

in 1969 at last had the opportunity to use the Atlas–Centaur payload to send more capable Mariners to Mars. The Jet Propulsion Laboratory in Pasadena, California, was able to design a craft with a weight of 413 kg and it was this payload that was launched twice, as Mariners 6 and 7, on 25 February and 27 March.

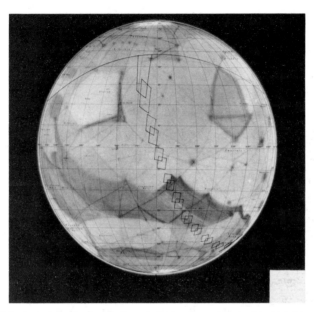

Even after Mariner 4 had revealed new facts about Mars this NASA map of the photographic coverage by the probe still includes a network of 'canals'. This shows how inadequate the coverage was until Mariner 9 in 1971 destroyed the myth of the canals for ever (NASA/JPL).

The single fairly simple camera of Mariner 4 was replaced by two TV cameras which were placed on a movable scan platform. The cameras took wide angle and telephoto pictures, respectively, and there were two tape recorders associated with them.

One was a digital tape recorder which was used to store the six least significant bits of an eight-bit digitally coded signal from every seventh picture element of each picture line. This was the most economic way of sending back picture data even at the new transmission rate, which

19

was 2000 times faster than Mariner 4.

An analogue tape recorder was used to store the analogue video signal from each pixel, and the two most significant bits of the eight-bit word were averaged over several lines and transmitted in real time. It was a far cry from the 8.33 BPS transmission rate of only four years before.

Mariner 6 arrived in the vicinity of Mars on 31 July 1969. Its success was rather overshadowed by the fact that only a few days before the crew of Apollo 11 had arrived back on Earth after their successful landing on the Moon. Nevertheless, there was an air of great anticipation at the JPL control centre as the close encounter approached. Those of us who were there will never forget the intense excitement as this new wave of planetary exploration began.

Far encounter pictures began returning from Mariner 6 54 h before closest approach. Between 48 and 28 h the narrow angle camera was used to obtain 33 full disc pictures of Mars, and they were transmitted over the next three hours.

From 22 to 7 h, 17 more far encounter pictures were recorded, and then, during 18 min of the close approach, during which Mariner 6 came as close as 3430 km to Mars, the two cameras took 25 pictures of the surface.

For the first time, the southern polar cap was imaged in some of the pictures. It was revealed as a sharply bounded area of ice or snow, with several craters showing up along the edge. One frame covering a total of 625 000 square kilometres of the surface was found to contain 156 craters varying from 3 to 240 km in diameter. In telephoto views many more craters down to 300 m in diameter were seen.

Mariner 6's scan platform carried both an infrared radiometer and ultraviolet and infrared spectrometers; these instruments were able to produce data on the constituents of the martian atmosphere for the first time. As expected, the tenuous atmosphere was largely composed of carbon dioxide, but it was also found to contain ionized carbon monoxide and atomic oxygen and hydrogen, although no nitrogen was detected.

A temperature as high as 16 °C was measured by the radiometer on the equator at noon, but on the night side of the planet the temperature could drop to as low as −73 °C.

Mariner 7 was on a slightly faster trajectory than its twin and made its closest approach to Mars on 5 August 1969. It took 93 far encounter pictures which appeared mottled and 33 at the near encounter.

The mottling turned out to be caused by craters with diameters up to hundreds of kilometres. The 'canali' of Schiaparelli, never absolutely disproved by telescopic observation, were not seen on the pictures. Discontinuous linear features and chance associations of craters and other markings may have deceived some observers. The study of climatology on Mars began with the Mariner 7 pictures, for some of them exhibited suggestions of clouds—a hint that was proved by later missions, which detected many clouds.

In the next phase of exploration of Mars, four explorers took off in May 1971, of which three reached Mars. Although Mariner 9 was the last of the three to leave Earth, it was the first to arrive, beating Mars 2 by 13 days and Mars 3 by 18. However, the launch window was not the occasion for complete triumph for America, for Mariner 9 was only half of a twin mission of which the other half, Mariner 8, failed. This time, it was the guidance system of the Centaur upper stage that failed after the Atlas booster had given the mission a good start. Mariner 8 did not reach Earth orbit but plunged to destruction in the Atlantic on 8 May 1971. Some urgent re-programming now had to be put in hand, for the two spacecraft had different roles in the exploration of Mars. Mariner 8 was to have been put into a polar orbit, covering the whole of the martian surface in the course of three months and returning up to 5400 pictures, while Mariner 9's orbit was to have been at an inclination of 50°, where it would be synchronized with the planet's rotation in such a way that its ground track would repeat, thus making any changes in surface features observable. In the frantic few days before the window closed, the Mariner 9 flight plan was modified to give a 65° inclination and a ground track repeating itself every 17 days. This was a way of maximizing the data and images returned now that the loss of Mariner 8 made the full mission impossible.

Only three weeks after the loss of Mariner 8, its sister ship took off from Cape Canaveral on 30 May 1971. This time the Atlas–Centaur combination worked perfectly and injected the craft towards Mars. The new explorer was based on the design for Mariners 6 and 7 but it had been extensively modified.

To begin with, it had been fitted with a retro-rocket using hypergolic fuels to enable it to be placed in orbit round Mars, and two large propellant tanks had been added for the nitrogen tetroxide and monomethyl hydrazine. Hypergolic fuels ignite on contact and therefore do not need an ignition system.

Next the camera systems had been upgraded. The wide angle camera was fitted with eight interchangeable filters, while there was a fixed yellow filter on the narrow angle camera.

21

There were five other experiments apart from the TV imaging system—an infrared spectrometer, an infrared radiometer, an ultra-violet spectrometer, S-band occultation and celestial mechanics. During the voyage to Mars the mass of the spacecraft was 975 kg and after the propellants had been spent in placing it into orbit round the planet this was reduced to 520 kg.

Mariner 9 travelled for 157 days, and on 14 November 1971, it became the first man-made artificial satellite of another planet. Its orbit round Mars was inclined at 64.28° to the equator and had a periapsis of 1397 km, an apoapsis of 17 916 km, and a period of about $12\frac{1}{2}$ h.

But the elation felt by the JPL team at Pasadena turned to dismay when the first television pictures began to come back and it was realized that the entire planet was covered by a huge dust storm. This had begun to be appreciated before Mariner 9 arrived through studies by Earth-based astronomers, but the true extent of the obscuration was seen only when the spacecraft's television system was turned on, some time before orbital insertion. Until the middle of December all that could be seen through the dust cloud was faint markings, and sometimes a diffuse feature followed by billowing dust clouds on its lee side. Pictures of the limb of the planet were analysed and showed that the dust reached an altitude of 70 km.

A T pattern of dark spots did appear on the early TV images and the theory that they were high spots on the martian surface showing through the dust proved to be correct when the dust cleared enough to reveal that they were in fact the summits of four enormous volcanoes. Each of the volcanoes covered an area as big as the state of Arizona and the biggest of them stood 29 km above the plain.

This was christened Olympus Mons and remains the outstanding feature on the surface of Mars. Olympus Mons is an immense shield volcano with a crater at its summit 70 km in diameter. Its base is 500 km across and it is instructive to compare its dimensions with those of Earth's largest volcano, the island of Hawaii, which is 200 km across on the ocean floor and rises to a height of 9 km above the seabed.

The other three volcanoes, Ascraeus Mons, Pavonis Mons and Arsia Mons, are only slightly less awesome than Olympus Mons and taken together show many of the features of terrestrial volcanoes. Erosion has been much less effective in the thin, dry atmosphere of Mars, however, and in some other respects the volcanic regions bear a resemblance to areas on the Moon.

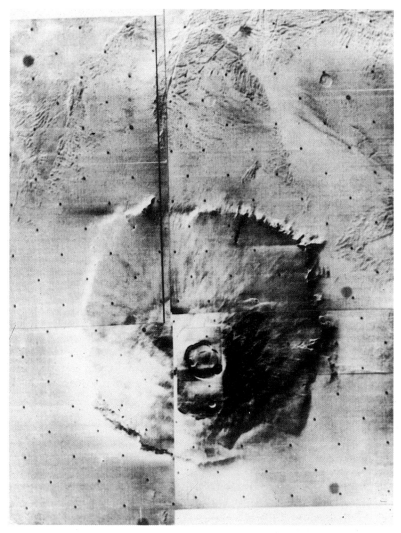

Olympus Mons was a big surprise to the science team who monitored the Mariner 9 pictures. It is the biggest known volcano in the Solar System, covering an area as large as France, and the central crater, a complex volcanic vent, is 65 km across (NASA/JPL).

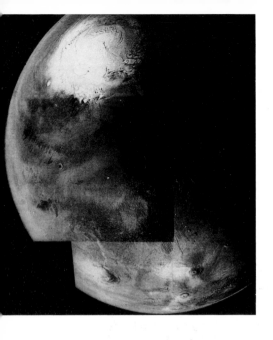

Much of the northern hemisphere of Mars is revealed in this mosaic of Mariner 9 pictures. The spiral patterning of the northern polar cap is at the top and the great volcanoes are the carbuncle-like protrusions at the bottom (NASA/JPL).

Previous Mariner flights had led to the erroneous conclusion that Mars was a dead planet, devoid of geological, or more correctly areological, activity. Mariner 9 disproved this theory and it now seems that the planet's entire surface may have been affected and even formed by such activity. Certainly, the extensive plains, pock-marked as they are by impact craters of all sizes down to the limits of resolution, were formed by lava flows and ash falls.

More than 7000 pictures were transmitted by Mariner 9, none being more spectacular than those that showed the huge system of canyons in the equatorial region.

This system is named Valles Marineris in honour of its discoverer. Valles Marineris is between 150 and 700 km wide and stretches for about 5000 km between 20 and 100 degrees west, roughly parallel to the martian equator. In terrestrial terms, it is as if a canyon up to 9 km deep stretched from New York to Los Angeles, or from London almost to Cape Town. Its origin largely remains a mystery. At the time of the discovery it was thought that the huge system had been cut by water. But reflection has brought the theory that, while it may have been enlarged by water, it was originally formed by tectonic movement of the planet's crust.

A minor side canyon in the system is roughly the same size as the Grand Canyon in Arizona. Great gulches up to 200 km wide stretch for thousands of kilometres and are up to 6 km deep. Other prominent features revealed by the pictures included immensely long faults and fractures in the martian crust, cliffs up to 4 km high stretching for hundreds of kilometres, chaotic landscapes of jumbled rocks, and many thousands of craters, both volcanic and formed by impacts.

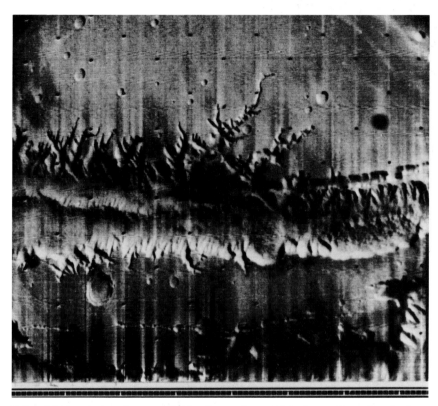

Valles Marineris is a huge gully in the surface of Mar's southern hemisphere. Subsidence along lines of weakness in the crust and possibly erosion by wind action are the probable causes of its appearance (NASA/JPL).

Unlike the Moon, Mars has a tenuous atmosphere and many of the Mariner 9 pictures show the effects of wind erosion and deposition on Mars. The equatorial region, it transpired, is mainly under the influence of erosion, the wind lifting dust and sand and scouring away hills

25

and cliffs and etching parallel grooves in flat areas. Nearer the poles, however, deposition of the wind-borne material dominates in the form of sand dunes and featureless spreads of dust.

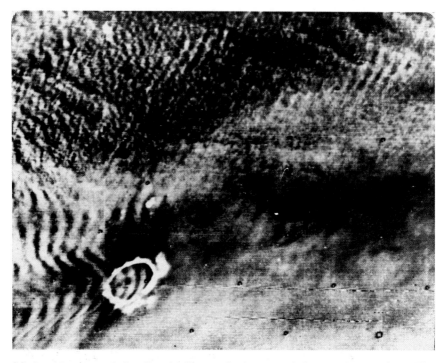

Mariner 9 returned the first really good pictures of the martian surface. This picture demonstrated that frost can appear on the surface, with the white-rimmed crater a prominent feature. Wave clouds are also visible (NASA/JPL).

Mariner 9 solved at least one martian mystery—the wave of darkening mentioned in Chapter 1. A tentative explanation before the space probes was that vegetation was responding to the release of water from the melting polar cap. But this was disproved by Mariner 9, which showed that the effect was an illusion.

This was not the only changeable set of features seen on Mars, which is by far the most volatile planet after the Earth. The polar caps of either carbon dioxide ice or water ice shrink and grow with the seasons and above all the pictures of martian clouds gave fascinating information to the meteorologists.

At the surface the martian atmosphere has the temperature and pressure of our atmosphere 30 to 40 km above the Earth. But even at these extremes carbon dioxide and water can freeze into crystals, so clouds do appear.

In the lee of large craters and other prominent features the phenomenon of lee waves was often seen. This is a type of cloud well known in mountainous regions of the Earth.

Wave clouds were detected by Mariner 9 in its 1971 encounter with Mars. These waves, with a wavelength of many kilometres, are caused by air flow over rough terrain (NASA/JPL).

The picture of Mars painted by Mariner 9 is one of a strange world if judged by terrestrial standards. The first human explorer to go there

27

will confront a world of high cliffs, immensely long and deep canyons, volcanoes of unbelievable size, dried river beds and jumbled rocks, all in a thin atmosphere where sound will scarcely carry. Reddish landscapes will shiver beneath pinkish skies decorated with occasional wispy clouds. He will have to be on his guard to keep away from landslides which may bring tens of kilometres of walls of canyon down in a huge cataract of rock. The temperatures will be so low that he will never be able to expose any part of his skin to the air, and he will always have to wear a pressurized spacesuit and rely on breathing oxygen from his backpack. It is not as strange as some of the fantasies about Mars created by science fiction writers and serious scientists alike, but it will be a truly alien world to the first human to land there.

One final aspect of the Mariner 9 mission was of great interest to astronomers. It was programmed cleverly to take close-up pictures of the two satellites of Mars, Phobos and Deimos. At least one well known astronomer had postulated that these little moons, which have unusual orbits round their parent, might be artificial.

Phobos completes an orbit round Mars in 7 h 39 min, and as this is much less than the rotation period of the planet, 24 h 37 min, the moon passes across the martian sky from west to east, instead of the usual east to west. From any part of Mars it is not seen for more than 3 h 10 min in any one rotation period.

Deimos, which is much further away, has a revolution period of 30 h 18 min. It travels through the sky from east to west at a rate of less than three degrees an hour.

The Mariner 9 pictures showed that they are irregular bodies, shaped rather like potatoes, and heavily cratered and natural, not artificial. As discussed in Chapter 4, Phobos in particular is the subject of active research from space.

The mystery of the 'canals' of Mars was, as mentioned above, cleared up when the first high resolution photographs of the planet began to return from space probes in the 1960s. They simply didn't exist outside the fertile brains of 19th century astronomers, aided by optical illusions in the poor seeing conditions of Earth-based telescopes.

But in a way the new pictures threw up a new mystery, for they showed that the surface of Mars is indeed liberally sprinkled with channels of all sorts of shapes and sizes. They range from what appear to be dried-up river beds which would be respectably sized wadis in a terrestrial desert area to the truly titanic Valles Marineris, which dwarfs anything seen on Earth.

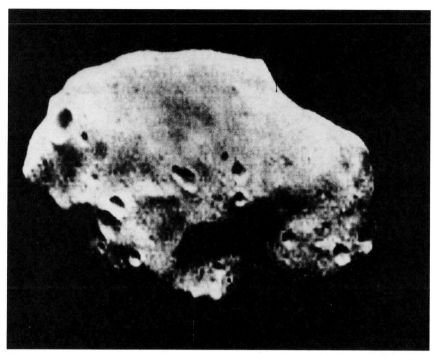

Mariner 9 was 5500 km from Phobos when this picture was taken in 1971. The moon is pock-marked with craters and is too small to have become spherical under its own gravitational force. It may be a captured asteroid and could provide much-needed volatiles for human explorers (NASA/JPL).

The interpretation of the Mariner and Viking pictures of the channels of Mars continues to this day and has generated some lively debates about their origins. The chief controversy is based on the fact that many of the channels appear to have been carved in the martian rocks by running water—on a planet where liquid water no longer exists on the surface.

The crucial questions to be answered, then, are 'Where did the water come from?' and 'Where did it go to?'

A subsidiary question is, of course, how liquid water could have existed on the martian surface for long enough—probably millions of years—to cause such enormous erosion effects.

As explained above, the atmosphere of Mars at the surface is about as dense as the Earth's atmosphere at an altitude of 40 km. The lower the atmospheric pressure, the lower the temperature at which water will boil and as a result liquid water would simply evaporate away in martian conditions as long as the atmospheric temperature was well above freezing; in addition, ice would sublimate.

The water vapour that results would rise high in the atmosphere and its molecules would be dissociated into their constituent oxygen and hydrogen atoms by the ultraviolet radiation from the Sun. Random motion of the atoms in the rarefied gas of the high atmosphere would in time lead to the entire stock of oxygen and hydrogen atoms reaching escape velocity, which is under half that of the Earth. It has been calculated, however, that the amount of water lost so far from the planet in this way would be sufficient to cover the surface of Mars to a depth of only 2.5 m. This compares with the 400 m of water which it is estimated existed early in the planet's life.

By far the greater part of the original water must have remained in place as subterranean ice and permafrost, perhaps covering and sealing in place reservoirs of liquid water deeper in the rocks. This, it will be seen, has an important bearing on the formation of many of the channels.

An enormous feature like Valles Marineris could not possibly be erosional in origin. The amount of material removed if by water erosion would have been so great that it would have been obvious in its new position as enormous fans of sedimentary material.

A close study of the valley system shows that it was in fact largely formed in a manner analogous to the sea floor spreading on Earth. More exactly, this is a tectonic feature formed on a huge dome which caused the martian crust to spread and crack. There is certainly some evidence that after the valley formed running water caused some erosion but it was not a major factor.

The westernmost area of Valles Marineris consists of a number of intersecting troughs caused by faulting, which has been named Noctis Labyrinthus.

Next are long narrow canyons called Tithonium Chasma and Ius Chasma, where landslides have produced heaps of debris on the canyon floors. These canyons are not well linked with one another and their floors do not make up a regularly graded slope, two reasons why they are not regarded as water erosion systems.

Some of these landslides, which have played a significant part in reshaping the surface of Mars, are bigger than anything of the sort seen on Earth. The biggest was probably in Ius Chasma, where about

30 km of the plain above the chasm slipped down along a front of 100 km. Some features that survive on Earth have resulted from landslides not too dissimilar in scale from those seen on Mars.

Such landslides will probably not be too great a hazard to human explorers, for the hundreds of slides that have occurred have probably taken place over many millions of years. Nevertheless, as a safety precaution, when planning landing sites the mission teams will probably rule out any locations which are close to the edges of canyons, either above or below the lip.

In the centre of the Valles Marineris system are wider canyons, Melas Chasma and Coprates Chasma, with straight scarps stretching for hundreds of kilometres. These have evidently formed from the collapse of the martian surface between faults, as the crater density inside the canyons is similar to that on the neighbouring plains.

Apart from the layered terrain at the poles, probably caused by climatic cycles, there are no obvious sedimentary rocks on Mars but these canyon walls exhibit layering in their upper sections, probably as a result of successive outflows of basalt in periods of volcanic activity. Geologists will be keen in the next century to get samples from these layers, for they will not only provide a record of the sequence of volcanic episodes on Mars, but they will also provide a means of dating these episodes by physical means, using ratios of radioactive isotopes.

In the east, the canyons widen out into an area of what is described as chaotic terrain, in which a depression is filled with a jumble of rocks and debris formed by continuous landsliding which deposits huge slabs of the rocks of the plains into the fault zones below. From the edges of this area channels extend out into Chryse Planitia, the area in which Viking 1 landed.

However, Valles Marineris is atypical; much more common are the valleys or depressions of various types which appear to owe their origins entirely to running fluid of some kind. They are usually sub-divided into three types in which water played a part and a fourth probably caused by lava, although there are so many different morphologies apparent that some defy a rigorous classification.

The 'broad channels' or 'outflow channels' are associated with the areas of chaotic terrain and are common around the Chryse Planitia. Generally speaking they flow from areas of chaotic terrain in the uplands, where craters are common, onto the smoother plains. They appear to have originated from episodes of catastrophic flooding and may be several hundred kilometres long, with a depth of more than a kilometre and a width of tens of kilometres, statistics which suggest erosion on an enormous scale.

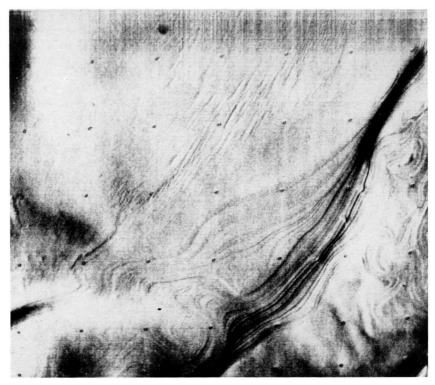

The swirling patterns of the terrain at the South pole of Mars are revealed in this Mariner 9 picture. The dark lines are terraces in the polar cap laid down in rhythmic fashion in changing climatic eras and this terrain will be of immense value in mapping the climatology of Mars when human explorers sample them (NASA/JPL).

It has been calculated that these run-off channels were formed by peak flood discharges of 10^7 to 10^9 m^3 s^{-1}. This compares with the average discharge of the Amazon of 10^5 m^3 s^{-1}. The only known flood on Earth that comes close to the martian discharge rate was the Late

Pleistocene flood when 'Lake Missoula' in Washington state in the USA burst through a glacial dam and discharged water at a rate of 10^7 m^3 s^{-1}, still only a hundredth of the possible maximum discharge of the martian run-off channels.

Not directly associated with chaotic terrain is a second type, channels which are narrow and sinuous and show a number of different forms. Thirdly, the cratered uplands exhibit a large number of small, closely spaced channels, also called run-off channels, which mark the sides of large craters and run along the plains between the craters.

Although many of the martian channels defy classification, there is indisputable evidence that many of the channels observed on Mars were formed or widened by water. Conditions on all the planets except the Earth are not favourable for the survival of water in liquid form, at least on the surface, but water ice is a common constituent of many of the satellites of the planets from Jupiter outwards and of the cores of the planets themselves.

It should therefore not be a surprise that so much water still exists on Mars, albeit in the form of sub-surface ice or aquifers. It has all been there since the formation of the planets more than 4 billion years ago and the main question to be answered is when did it appear on the surface and make its contribution to the geomorphology of Mars.

Several scientists have concluded that the channels have an extremely ancient provenance. None is younger than 100 million years old and the origin of many stretches back almost 4 billion years, close to the formation of Mars. Since future exploration and perhaps settlement of Mars may depend on the easy availability of water, the mere fact that the channels are so ancient should not be taken to mean that the water that formed them is no longer present. It has been estimated that 1 kg of water can be extracted from 100 kg of martian soil simply by heating it, since it contains hydrated minerals.

If this turned out to be a difficult process, a simple solar-powered compressor could be used to compress the tenuous atmosphere. By merely doubling the pressure it would be posssible to extract by condensation 453 cm^3 of water from 28 000 m^3 of atmosphere (that is, 1 lb of water from a 100 ft cube of atmosphere). There is certainly not going to be a shortage of water on Mars.

3   The Search for Life

How remarkable! We are performing chemical and biological experiments as though in our own laboratories. Taking pictures at will, listening for seismic shocks and making measurements of the atmosphere and surface. All of this from the first spacecraft ever to be landed successfully on Mars.
Gerald A Soffen

In the history of the exploration of Mars the Viking project will always command an unrivalled place. It is true to say that the success of the project led to an accretion of knowledge of Mars which will still be used as a baseline when humans are landing there in the next century.

Vikings 1 and 2 were highly sophisticated craft, each with an automatic laboratory capable of detecting chemical activity associated with the sort of life we know. But the Viking project also embraced wide-ranging meteorological, geological, chemical and seismological studies, as well as a mapping programme that eventually photo-graphed almost the entire surface of Mars.

Like all the other American interplanetary missions, Project Viking was launched from Cape Canaveral. The launch vehicle was the powerful Titan–Centaur, a 631 tonne rocket that had begun life as an ICBM. The Vikings began their 10 month and 11 month voyages to Mars on 20 August and 9 September 1975, respectively.

On 19 June 1976, the retro-rocket of Viking 1 was fired to put the whole spacecraft—orbiter as well as lander—into a highly elliptical orbit around Mars. Periapsis was 1500 km and apoapsis 50 600 km, while the inclination to the martian equator was 37.8°. It transpired that the chosen landing site was too rough, and many hours of discussion and examination of radar data and hundreds of pictures taken by the orbiter followed. In the end it was on 20 July 1976 that the historic landing took place. The landing site finally chosen as relatively free from the hazards of large rocks or deep depressions was in the plain known as Chryse Planitia, at a point 22.27° north of the equator and at a longitude of 47.97°. (Longitude on Mars is measured from an arbitrary meridian agreed on by international astronomers.)

Viking 1 was 350 million kilometres from Earth when the lander separated from its 'mother' craft. Seven minutes later a compact computer in the lander issued a series of commands to small rocket thrusters on the heat shield. The motors fired for more than 22 min to reduce the lander's velocity and throw it out of its orbit around Mars. Two radar altimeters fed altitude and velocity information to the computer which had to be virtually autonomous at this stage since commands from the control centre at the JPL would have taken 19 min to reach the spacecraft.

Eventually, the instruments in the lander began to sense the fringe of the martian atmosphere and 3 min 59 s later, when the craft was 6 km above the surface, a big red and white parachute was deployed. This slowed down the lander from 830 to 188 km h^{-1} and seven seconds later the heat shield was jettisoned. The descent was cushioned for the last few metres by small rocket motors and the Viking 1 lander touched down in a dusty, rock-strewn plain.

Minutes after touchdown the lander's television camera relayed back to Earth the very first picture ever taken from the surface of Mars—a shot of its own footpad resting on rocky martian soil.

From Chryse Planitia the Viking 1 lander transmitted images of a fairly flat but undulating landscape of ochre-coloured material, strewn with boulders lying between small dunes. The dunes were, of course, of windborne material and other wind effects were obvious. They included trails of fine grains lying between the boulders, carried there by prevailing winds in quite consistent directions. The entire landscape did not, however, owe its morphology to erosion and deposition by winds. There were visible stretches of a hardened vitrified surface, demonstrating a process of water evaporation leaving behind a crust of mineral salts.

This was a landscape not unlike the stony deserts of North Africa, North America and Asia though not, of course, with the hardy desert plants to be found on Earth.

One of the bizarre aspects of the first colour picture, which was received the following day, was that in processing it with the aid of computers the scientists at JPL had a certain amount of scope in altering the colours perceived. At first, basing their view on Earth conditions, they made the sky light blue. This version of the picture was released to the avid networks within eight hours of the image being received at Pasadena and was widely shown on television. But later, after more thought, JPL adjusted the contrast so that the genuine colour of the martian sky, pink, was substituted for the blue. When

An unpromising scene for human settlers from Chryse Planitia, where Viking 1 landed. However, human endeavours may alter the face of the planet over a long period of years (NASA/JPL).

this was announced by James Pollack, a member of the imaging team, at a press conference it was greeted with booing.

This was blamed by the JPL scientists on 'Earth chauvinism.' Carl Sagan, another member of the team, commented 'The sort of boos given to James Pollack's pronouncement about a pink sky reflects our wish for Mars to be just like Earth.' The pinkness of the sky was actually extremely interesting, for it revealed that reddish dust from the surface is always present in the atmosphere; this must result in a slightly higher temperature than would otherwise be the case, for the dust absorbs heat from sunlight.

Viking 2 entered a 1502–35728 km orbit around Mars on 7 August 1976; this time the inclination of the orbit was 55.6°. A site at 44 °N had been chosen but Viking 1 pictures showed that this site was situated in extremely severe terrain. In the next few weeks no less than 4.5 million square kilometres of the martian surface were scrutinized before a new site was chosen.

As a result, on 3 September 1976, the Viking 2 lander was set down some 7500 km from Viking 1 in the area known as Utopia Planitia. The coordinates of this site were 47.96 °N latitude and 225.77° longitude.

The landing team had chosen this site because pictures from the orbiters had seemed to suggest that the area was safer because it was covered with sand dunes. As it happened, the lander found a stony plain stretching in all directions and there was no sign of dunes. The larger boulders and pebbles in the field of view may have originated as debris from a meteoric impact crater about 200 km away. Between the rounded boulders, which have a spongy, vesicular appearance typical of some terrestrial lavas, appears a polygonal pattern of channels.

Viking was a large craft compared with previous American interplanetary probes. The orbiter had a mass of 2325 kg before insertion into Mars orbit, and 950 kg after the fuel was exhausted.

The lander had a mass of 1090 kg on being ejected from the orbiter, and 600 kg after its rocket had fired. On board the Viking 1 lander the computer had been pre-programmed to dig a trench with the aid of the arm which was part of the sampling equipment, but the first pictures from the landing site showed that the chosen point was too rocky. So the computer was sent a signal to re-position the digging arm and the unique task of sampling martian soil began soon afterwards.

There were three Viking experiments which were aimed at finding the existence of life or its remains on Mars. The first was the pair of cameras mounted on the lander to take black and white, colour and stereo pictures. They would have pictured any large life forms or plant growth, but they failed to do so. Next, the GCMS (gas chromatograph/mass spectrometer) searched for organic molecules in the soil. Carbon, hydrogen, nitrogen and oxygen in the form of organic compounds are present in all living matter on Earth. The GCMS searched for these compounds either as evidence of life, its precursors or its remains but found none.

Organic molecules were expected to be present in the martian surface strata even if there was no evidence of existing or fossil life, because organic materials are found in meteorites and they must be deposited on the martian surface as they are on Earth. The conclusion is that some process on Mars actually destroys organic molecules—perhaps the ultraviolet flux from the Sun does this.

Finally, there was the biology experiment. This consisted of a small box containing three instruments which JPL described as the most sophisticated scientific hardware ever built. The objective of the instruments was to search for signs of metabolic processes like those used by bacteria, green plants and animals.

37

In one section, the gas exchange experiment, martian soil was exposed to water vapour for six martian days and then wetted with a complex solution of metabolites which would act as a nutrient to bacteria. The controlled atmosphere of helium, krypton and carbon dioxide above the soil was then monitored by gas chromatography for any signs of respiration by organisms. Various amounts of oxygen, nitrogen and carbon dioxide were detected by sensitive instruments, depending on the amount of nutrients supplied to the soil sample.

However, this was not taken as a sign of life, for all the changes could be explained purely in terms of chemical reactions.

In the pyrolytic release experiment, a tiny sample of soil was exposed to carbon monoxide and carbon dioxide containing radioactive carbon-14 in the presence of light to see whether organisms employing photosynthesis were present or whether the process known as chemotrophy, in which the soil and water vapour reacted together, had mimicked it.

On Earth, plants grow by photosynthesis and in doing so they incorporate carbon from atmospheric carbon dioxide into their cells and expire oxygen. If martian organisms did the same, it was reasoned, some of the carbon-14 would be detected in the soil sample.

At the end of the exposure the radioactive gas was purged from the experimental chamber and the sample was heated to drive off the carbon-14. In some of the trials of the pyrolytic release experiment a surprising amount of carbon-14 was found to have been incorporated into the samples, but it was decided that this was the result of exotic chemistry rather than photosynthesis.

Finally, in the labelled release experiment soil was moistened with a solution containing several organic compounds, nicknamed 'chicken soup' by the scientists, which was labelled with radiocarbon. Radioactive gas was released after an exposure of a few days, as would be expected if metabolism was taking place. *Prima facie* this was evidence of life processes, but there were some puzzles to be solved.

There was a big difference between the amount of gas given off by unsterilized and sterilized samples, just as there is with samples of Earth soil, which invariably includes bacteria and other organisms. One interpretation of the results is that unsterilized soil contained organisms and gave off carbon dioxide, while sterilized samples gave off very little because the organisms were killed.

However, unsterilized soil samples which were given a second injection of nutrient did not produce large quantities of gas. The orthodox conclusion is that the existence of life in the martian soil was not proved, although not all scientists believed this to be proved beyond doubt.

Perhaps the release of the gas from the soil indicated that it contained oxidants, which on Earth perform the role of breaking down organic matter and living tissue. One view is that they exist in the form of peroxides or superoxides which fizz and froth in the presence of water.

The conditions now known to exist on or just under the surface of Mars do not allow carbon-based organisms to exist and function. If life has evolved on Mars it may continue to exist in isolated 'oases' where conditions are particularly good, such as near the equator and in deep depressions. The Viking samples came from two tiny sites on opposite sides of the planet and can hardly be representative.

On the other hand, the question as to whether life in the distant past existed is still open, for the pictures from Mariner 9 and the even more extensive coverage of the Viking orbiters revealed that liquid water did exist on Mars in large quantities in the past.

It is obvious that the debate about the existence or non-existence of life on Mars will continue to rage until samples are taken from many different areas of the planet's surface, including beneath the surface.

Analysis of the soil by the two landers showed that silicon and iron are the most abundant elements. About 45 per cent of the soil consists of silicon dioxide and 19 per cent iron oxide, much of it in the form of maghemite, which undoubtedly gives the planet its characteristic red colour. Other elements present are magnesium, calcium, sulphur, aluminium, chlorine and titanium. There is 100 times as much sulphur as is found in terrestrial soil, but there is much less potassium.

The high iron and low potassium have profound implications for the geology of Mars, which has probably not suffered as much differentiation of the elements by internal heating as the Earth has. Earth has a large core consisting of iron and nickel which have sunk into the interior after being heated as the planet accreted.

The general conclusion from the Viking studies was that the martian soil might be a mixture of argillaceous minerals high in iron and iron hydroxides, together with other minerals high in iron and carbon. About one per cent of the soil by weight was found to be water.

The martian surface had been expected to be a dangerous place for the Viking landers, but in fact only minor changes were recorded by the cameras. There were slight variations in the brightness and colour of a few places on the surface where dust was moved in thin layers and a couple of tiny landslips near Lander 1.

The landers made one momentous discovery about the atmosphere—it contains 2.7 per cent of nitrogen, one of the essential elements of life. No previous spacecraft had detected nitrogen in the predominantly carbon dioxide atmosphere.

A seismometer on Lander 2 found no evidence of quakes on Mars, but meteorology instruments measured air temperatures and wind speeds and directions. They were able, in fact, to compile the first extra-terrestrial weather reports in the history of meteorology. This was how Seymour L Hess, meteorology team leader, reported on conditions at Chryse Planitia on the second and third 'sol' (martian day).

> Winds in the late afternoon were again out of a generally easterly direction but southerly components appeared that had not been seen before. Once again the winds went to the southwesterly after midnight and oscillated about that direction through what appears to be two cycles. The data ended at 2.17 PM (local martian time) with the wind from ESE, instead of from the W as had been seen before. The maximum mean wind speed was 7.9 metres per second but gusts were detected reaching 14.5 metres per second.
>
> The minimum temperature attained just after dawn was almost the same as on the previous sol, namely −86°. The maximum measured temperature at 2.16 PM was −33°. This was 2° cooler than measured at the same time on the previous sol.
>
> The mean pressure was 7.63 mb, which is slightly lower than previously. It appears that pressure varies during a sol, being about 0.1 mb higher around 2 AM and 0.1 mb lower around 4 PM.

It appears that there is very little change in the weather on Mars from day to day; unlike the Earth Mars does not have large oceans and huge quantities of atmospheric water vapour, which cause our weather extremes.

Lander 2 functioned until 11 April 1980, but Lander 1 carried on working, sending back weather reports and pictures, until November 1982.

While the landers were attracting the majority of the attention, especially because of their search for life, the orbiters were conducting their own programme of research. In addition to their cameras, with which new and startling pictures of the variegated landscapes of Mars were captured, the orbiters carried an atmospheric water detector which mapped the atmosphere for water vapour, an infrared thermal mapper which measured temperatures, including seasonal changes, and their own radios which were used to measure the density of the atmosphere by the distortion of the signals as they passed through it.

The thermal mapper provided new information about the polar caps. When Mariner 4 returned a temperature of $-62\,°C$ for the polar caps it was concluded that they contained frozen carbon dioxide, or 'dry ice'. The temperature was simply too low for water ice to be present. Viking confirmed that this was true of the southern pole, but in the north its measurements showed that the carbon dioxide ice is a winter phenomenon and that in the summer the polar cap does consist of water ice, the CO_2 having returned to the atmosphere. This discovery alone was almost enough to justify the project, for it radically altered views about the availability of water on the surface of Mars and may have a profound effect on plans for manned missions.

The four telephoto lens cameras on board the two orbiters transmitted more than 51 000 images to the JPL control room through the Deep Space Network over the years. The pictures gave a complete picture of the topology of Mars, showing that most of the southern hemisphere is higher than the average surface level, while most of the northern is lower. There are exceptions: the large, deep basins in the south known as Argyre and Hellas lie far below the mean surface. The northern hemisphere has three raised areas: the Syrtis Major Planitia, the Elysium Mons volcanic province and most notable of all the Tharsis ridge.

This gigantic 'continent' measures 2500 by 1900 km (1500 by 1200 miles) and includes the mammoth Olympus Mons and three other volcanoes first observed by the Mariner 9 cameras. The ridge covers a quarter of the entire martian surface and its bulge probably began to rise 3.3 to 4.1 billion years ago.

Another highly significant feature of the martian surface detected by the Orbiter cameras was the layered terrain at the north polar cap, evidence of climatic change in the past. In addition, many new images of the Valles Marineris and other channels were received.

The Viking imaging team reported that on the arrival of the craft before the summer solstice in the northern hemisphere, the hemisphere was covered by a relatively dense haze, while the south was much clearer. When spring came to the south, the southern hemisphere was covered by an even denser haze and even when this cleared much of the south was obscured by dust storms.

Meanwhile, northern latitudes were obscured by condensate clouds and hazes during the autumn and winter. North of $60\,°N$ a featureless 'polar hood', believed to consist at least partly of particles of CO_2 ice, obscured much of the surface detail.

Fronts swept the zone between 40 and $60\,°N$, moving south out of

41

the polar regions and producing clearly visible clouds. Unlike fronts on Earth, of course, the fronts did not result in rain.

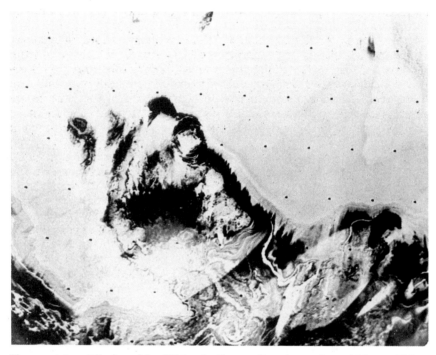

The great ice cliffs found by Viking in the north polar region of Mars will be a prime target for investigation by the first explorers. They may reveal the fossil history of the martian climate and will also be a major source of water (NASA/JPL).

Haze extended to between 35 and 40 km above the surface, as shown by a number of pictures of the planet's limb.

Clouds form around the giant volcanoes of Mars in the late morning because the atmosphere is forced to great altitude by the huge volcanic shield and in doing so is cooled, and they obscure the flanks of the volcanoes to a height of 20 km.

This cleared up a long-standing mystery because the clouds characteristically take up a 'W' shape because of the topography of the volcanoes, which has long been observed telescopically from Earth.

The observers, of course, had no way of knowing how the clouds were formed because the existence of the volcanoes was unknown before Mariner 9. One popular interpretation at the beginning of the century, as mentioned in Chapter 1, was that the Martians were preparing for war.

Wave clouds, with a wavelength of about 60 km, appear in a mosaic of an area north of Olympus Mons. A large crater with a raised rim, Milankovic, lies in the path of strong westerly winds and causes the waves by perturbing the airstream. Waves are common on Mars, and this provides useful information as to the stability of the atmosphere. The fact that the waves sometimes persist for hundreds of kilometres shows that the atmosphere is highly stable, otherwise the waves would be quickly destroyed by turbulence.

Some other weather phenomena were recorded by Viking. In the bottoms of some craters and channels Orbiter pictures showed that early morning fog consisting of ice crystals formed. Apparently in the morning a small amount of water vapour is driven off the surface by the Sun and recondenses in the colder atmosphere to form ice crystals. Although the amount of water is tiny—if liquefied it would form a layer about a micron deep—this phenomenon may have implications for future biological exploration.

A low-pressure cell similar in appearance to an extratropical cyclone on Earth, complete with counterclockwise circulation, was seen in an area near the North Pole. The passage of cold fronts, in which warm, moister air is lifted over a wedge of colder, denser air, was observed many times. As the air rises water forms ice crystals, so making the fronts visible through the formation of clouds. The result of all these observations was the formation of a much clearer view of the weather phenomena on Mars which added to the data available for the new science of comparative planetology. Greater knowledge of climatic conditions on other planets should enable us to understand terrestrial weather more clearly.

All the systems were still in operation when Orbiter 2 finally ran out of attitude control gas on 25 July 1978. With no attitude control it was impossible to point the spacecraft at the targets, so the mission came to an end on that date.

Orbiter 1, however, soldiered on beyond the end of the seventies and was finally declared dead only in the late summer of 1980. It was the end of an historic era in planetary exploration which had lasted 18 years and which will undoubtedly be matched but not surpassed in its importance in the 1990s.

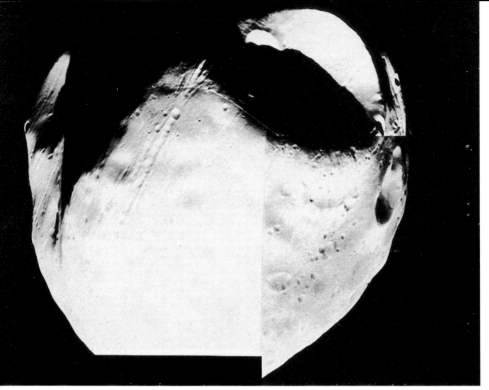

The crater Stickney dominates this view of the martian moon Phobos. With a diameter of 10 km it must have come close to shattering Phobos and the moon still bears cracks many kilometres long which may have resulted from the impact (NASA/JPL).

4   The New Generation

The flight of two Soviet spacecraft towards Phobos will become the first stage of a vast Soviet programme for the exploration of Mars in the coming decades.
TASS

It is a strange fact of history that no craft were launched towards Mars between 1975, the year of the two Vikings, and 1988, when the Russians began their new campaign of exploration. From the American point of view this was partly because of budgetary restraints and partly because the decision had been taken to use the much-delayed Space Shuttle for all future planetary launches. The Soviet planners had, on the other hand, become totally preoccupied with missions to Venus and to Halley's Comet. Even their enormous commitment to planetary exploration could not cope with a Mars programme as well.

But now that the Soviet Union appears to have exhausted all the possible options for fruitful missions to Venus, its vast resources in manpower and space technology are apparently going to be focused on Mars. Several types of mission have been described by Roald Sagdeyev, the former director of the Space Research Institute of the Soviet Academy of Sciences, and other officials.

Even if not all of them actually fly, the dual Phobos mission which began in July 1988 is sufficiently ambitious to be worthy of special mention.

It surpasses in complexity all past Soviet planetary missions, and complete success with Phobos would be equivalent to anything that the Americans have done, and it is unfortunate that this success was compromised early in the mission when Phobos 1 was accidentally 'switched off'.

Paradoxically, it is the expertise which the Americans have built up in communicating with spacecraft far from the Earth which it is expected will enable the Soviet planners to get full value from the remaining Phobos.

NASA has three dish antennas, each 70 m in diameter, at Goldstone in the California desert and in Spain and Australia. They are positioned round the globe so that a planetary spacecraft is always in 'view' from at least one of them. This Deep Space Network (DSN), developed by the JPL over three decades, can pinpoint the position of a distant craft to within a few metres and its velocity with similar precision. During the Phobos missions it is acting in concert with the Soviet Union's own tracking network, centred on a multiple array of antennas at Yevpatoriya in the Crimea.

Tracking with the precision of the DSN was needed for the final phase of the mission to the moon Phobos, and to obtain this accuracy the JPL team used the radio astronomy technique known as very long baseline interferometry (VLBI). Widely spaced antennas on Earth were paired to obtain measurements more accurate than one antenna could achieve. The DSN received telemetry from the one operational craft, including images and instrument readings, but its principal responsibility was to be the ranging and VLBI measurements.

The moon Phobos rotates in a period of 7 h 37 min, so the antennas on the landers would not be pointing at the Earth all the time. The DSN expected to pick up signals for only 17 min for each rotation period, but without its special facilities this transmission period would be even shorter.

The Mars opposition of autumn 1988 was among the more favourable in the fifteen year cycle, so the Proton rockets used in the Phobos project were able to place a large payload into low Earth orbit (LEO) and then on to the trans-Mars trajectory. The objectives and hardware that had emerged from the relative secrecy that still surrounds the Soviet space programme would have been regarded as remarkable even if they had been the product of the Jet Propulsion Laboratory.

Each craft has a mass of about 7 tonnes, close to the maximum for such a flight. Such are the advances made by the Soviet design teams since their first unmanned Mars probes that they have progressed to a highly sophisticated and densely instrumented craft that has an amazingly ambitious programme.

The Phobos spacecraft is a new class of Soviet design, with two square solar panels protruding from the sides of a central body which is crowned by a high-gain antenna. The mission plan was originally for each craft to go into a highly elliptical orbit of low inclination round Mars for about two months, but this was obviously modified after the loss of Phobos 1. Phobos 2 did go into orbit on 29 January 1989.

Then the intention was to circularize the orbit by small propulsive bursts to match the orbit of Phobos at a distance of about 9400 km from Mars. During the early stages of the mission television cameras were programmed to image large areas of the planet's surface and radio-meters and photometers on board were to study changes in soil temperatures during the day and the seasons. They will also be looking for permafrost and for areas where internal heat is reaching the surface, suspected from some Viking images.

The solar cell panels extend like wings from the side of this model of the Phobos spacecraft. Most of the scientific instruments are in the top half of the craft, above the spherical propellant tanks (author).

Gamma radiation studies will further enhance knowledge of the principal elements in the martian rocks—mainly aluminium, magne-sium, sulphur and iron. Natural radioactive elements will be measured to help clarify certain points in the thermal history of Mars.

The study of the martian atmosphere and ionosphere will also be continued by the Phobos probe. Spectra will be recorded of solar radiation passing through the atmosphere as the probes go in and out of the planet's shadow. In this way it should be possible to calculate the altitudinal distribution of ozone, water vapour, molecular oxygen and dust and also to record daily and seasonal variations in the temperature and pressure and other atmospheric characteristics. Measurements will also be made of the deuterium/hydrogen ratio in an attempt to find out where the liquid water has gone.

Another important study is that of the martian magnetism. Previous missions have established that if there is a magnetic field around the planet at all it is tens of thousands of times weaker than the Earth's. The strength of a planet's magnetic field is an important clue to the internal structure—for instance, the Earth's field is so strong because of its metallic core—and the weakness of the martian field indicates that any core that does exist cannot be more than one per cent of the planet's mass. Thus, it can be shown that the early history of the two planets was different; the Earth being so much more massive has suffered more differentiation, with the heavy iron and nickel penetrating to the core.

The Phobos 2 craft, carrying no fewer than 31 experiments, will also carry out studies of the magnetosphere of Mars and its interaction with the solar wind. Even during the cruise phase between the Earth and Mars the instruments were programmed to observe the Sun's corona and upper atmosphere, solar flares and solar oscillations, and to measure cosmic rays and solar gamma-ray bursts.

But the name of the mission is a clue to its major purpose and that is a unique investigation of the planet's major moon. Phobos, together with Deimos, was photographed at close range for the first time by the Mariner 9 spacecraft back in 1971 and 1972, revealing them to be heavily cratered non-spherical bodies with some fascinating surface features.

Higher resolution pictures from the Viking orbiters in the mid-seventies gave more details. The satellites are both locked into stable synchronous rotation about Mars, with their longer axes pointing towards Mars and their shorter axes normal to the planes of their orbits. This means that Phobos rotates round its shorter axis in the same time that it revolves around Mars, pointing the same face towards its parent.

The albedo of about 0.06—meaning that only 6 per cent of the sunlight falling on Phobos is reflected—has been produced by the constant bombardment of the moonlet in early times by meteorites which have totally shattered the surface so that the regolith is hundreds of metres deep. The surface of Phobos is dominated by sharp, fresh-looking craters of all sizes and a vast network of linear features resembling crater chains.

These grooves, tens of kilometres long and hundreds of metres wide, appear to be surface fractures associated with the crater known as Stickney which at 10 km in diameter is not large by martian standards but dominates a body which only measures 21 km by 19 km. Crater densities on both satellites are comparable to that on lunar uplands, a fact that suggests ages of a few billion years.

The density of Phobos was estimated to be about $2\,g\,cm^{-3}$, much the same as that of carbonaceous chondrite meteorites, evidence that it may have been a captured asteroid and an important pointer to its possible usefulness as a source of raw materials when manned expeditions are mounted.

Thus it is a fascinating body which may well be the first outside the Earth–Moon system on which men will land. This makes the Soviet Phobos mission particularly interesting.

The actual investigation of Phobos will proceed in several stages. In one, a laser called Lima, suspended beneath the spacecraft, will be fired in 10 nanosecond pulses from a range of 50 m. The resultant energy will vaporize and ionize the particles of soil the laser beam strikes to a depth of between 1 and 2 microns. Some of the scattered ions will reach the spacecraft, where a mass spectrometer will be able to detect the elements and isotopes which the soil contains. Each cycle of measurements is expected to register about a million particles and there will be about 100 such cycles during the 20 minute encounter with Phobos.

A pulsed ion beam will be used in another experiment, which has been given the name Dion. It will fire ions of the inert element krypton at the soil of Phobos, knocking ions out.

These too will bounce up to the spacecraft, where a sensitive mass spectrometer will analyse them, looking in particular for ions that might have been implanted by the solar wind.

The spacecraft are equipped with radar sets which will direct pulses at the moon at three different frequencies, penetrating to a depth of 200 metres. The aim of this is to gather information on the electrophysical properties of the regolith and perhaps to prove the theory that the surface has been shattered to this depth.

Meanwhile, a lander will be detached from the main body of the spacecraft and will be propelled the few metres to the surface, where a pyrotechnic charge will fire a spike into the soil to anchor it. Solar panels will then open on the lander to draw power from the Sun and observations will begin. One result should be an unprecedented level of accuracy in measurements of the orbit of Phobos, for the control

49

team back on the Earth will be able to use radio signals from the main spacecraft and the lander in the VLBI interferometry mode to pinpoint the position of the moon to within 5 metres.

It is believed that there will be internal movements in Phobos caused by the tidal forces generated by Mars, and a seismometer will record their magnitude, and will also measure tremors caused by the thermal expansion of the rock at sunrise and sunset and meteorite strikes.

Even the impact of the lander on the soil will be used as a means of measuring the physical and mechanical properties of the surface rock, and an x-ray fluorescence spectrometer will provide another means of analysing the soil. In addition, another television camera is installed on the lander to send back views of the surface at three wavelengths so that colour pictures can be constructed in the computer labs back on Earth.

More details of the soil characteristics will be provided by an ingenious second lander which will not be fixed. It is in the form of a sphere with attached springs which will enable it to 'hop' across the surface, sending back measurements of the soil's characteristics. Each leap will carry it 20 metres and there will probably be about ten leaps before the batteries are exhausted.

There are also some fundamental questions about the Sun which may be answered during the voyage to Mars, during which the angle formed by the Sun, the Earth and Mars grew from 0 to 180°. This, according to the Soviet scientists, created an excellent opportunity for simultaneous observations of the Sun from Earth, from artificial satellites of the Earth, and from the Phobos probes. It provided a unique chance for modelling the three-dimensional structure of the Sun's chromosphere and corona.

The Soviet manned space programme has always been particularly sensitive to possible radiation damage from solar events and this aspect of the Phobos missions will give them added confidence that their forecasting of anomalous solar activity is accurate.

They believe that the images of the Sun in soft and ultra-soft x-ray frequencies and in visible light will help them analyse physical conditions inside flares, coronal gaps and bright spots and get a better understanding of the transformation of magnetic energy into thermal waves which heat up coronal plasma.

Project Phobos is not purely a Soviet affair, however. Organizations, experimenters and specialists from Austria, Bulgaria, Hungary, East Germany, Poland, Finland, the Irish Republic, France, West Germany, Czechoslovakia, Switzerland, Sweden and the European Space Agency have all taken part, and there has been some input from NASA as well.

The Phobos spacecraft is not a one-off design. Unlike some of the expensive American interplanetary craft of the 1960s and 1970s it was not specifically designed for this mission, but is what the Americans call a 'bus'. It consists of a basic spacecraft onto which the various instruments, probes, landers and sub-systems needed for particular missions can be mounted. NASA is now adopting a similar philosophy for its Planetary Observer series of spacecraft as it saves money.

Phobos derivatives, Soviet officials have said, will be used for missions to both Venus and the Moon, as well as other flights to Mars.

A propulsion system used to make mid-course corrections and to provide the burns putting the spacecraft into martian orbit will be jettisoned once this stage is accomplished. This will leave the second stage of the combined braking/correction propulsion system (CBPS), which employs 28 small hydrazine thrusters to carry out the manoeuvres around Mars and Phobos. Earlier Soviet spacecraft were not always particularly sophisticated from the control point of view, but Phobos is equipped with a capable central digital computer complex which is said to have a memory capacity of 30×10^{16} bits. Many aspects of the operations of the craft will be controlled autonomously by the computer, since the return communications delay time between Earth and Mars will be anything up to 40 minutes. This will make it impossible for the controllers on Earth to supervise all activities.

At the time of writing no results have been announced from the Phobos craft but it is legitimate to assume that extremely valuable scientific data will be gained at the end of the 460 days of the nominal missions. These results will be useful to the Soviet planners who have already begun to build craft which will be going to Mars in the 1990s.

The Soviet Union, although it is throwing immense resources behind the present campaign, will not be returning to Mars until the 1994 launch window. This means missing two launch windows, but the 1994 mission promises to be extremely fruitful. Stepping up to an even higher level of complexity, the planners at the Babakin Design Bureau hope to deploy a sort of second generation Viking. There will be a satellite in orbit around Mars, a lander on the surface to carry out physical, chemical and biological research and, in addition, a balloon which will spend a protracted period travelling over the martian surface.

This is based on a concept which is also being explored by American scientists. A balloon filled with helium will rise in the martian day as

51

the Sun warms the helium and makes it expand. Then, as the Sun sets, the balloon will lose volume again and descend towards the surface. In this way telemetry on the nature of the atmosphere and its changes over the diurnal pattern and over a more extended period can be transmitted to the orbiter for relaying to Earth.

American scientists working on this idea have tested a sophisticated helium device which will fly at an extremely low altitude and trail an instrumented 'tail' over the surface to take samples and analyse them.

According to Roald Sagdeyev, the maximum payload version of the 1994 mission would also include a small rover, surface probes and penetrators and small weather stations. The rover, now being developed, would be capable of travelling several tens of kilometres and perhaps more than 100 km, depending on surface conditions.

It would look similar to the American design illustrated, consisting of an articulated vehicle with six metal wheels enabling it to negotiate gullys by bending in the middle, with the middle pair of wheels going

The six wheels of an automatic rover negotiate the rough terrain of the martian surface while a launcher waits in the background to receive the soil samples it collects. After rendezvous with the Earth return stage in orbit around Mars the sample will be returned to a quarantine facility in orbit around the Earth (NASA).

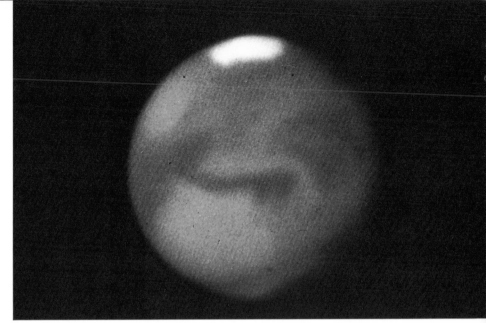

Even with the best Earth-based telescopes it is difficult to discern much detail on Mars. This picture, by Dr Robert Leighton, reveals a bright polar cap and darker patches around the Equator, but the true meaning of the markings became clear only with exploration by spacecraft (NASA).

The Mariner 9 spacecraft, the most sophisticated produced by the Jet Propulsion Laboratory to that date, completely revolutionized man's view of Mars. The pictures it sent back showed that it was a much more interesting planet than was previously thought (NASA/JPL).

Lift-off from Cape Canaveral. One of the two Viking spacecraft begins its journey to Mars on top of the powerful Titan-Centaur launching rocket (NASA/JPL).

A Viking lander as it would appear on the surface of Mars to the first human explorer who will seek it out. When eventually it is retrieved it will be thoroughly searched for any terrestrial organisms that may have survived (NASA/JPL).

Viking 1 Orbiter took this striking picture of Mars with half the globe illuminated. Valles Marineris can be seen in the left-hand lower corner.

The first colour picture to be returned from Chryse Planitia by Viking 1's lander showed an undulating landscape of ochre-coloured material strewn with boulders lying between small dunes. The big surprise was the pink sky (NASA/JPL).

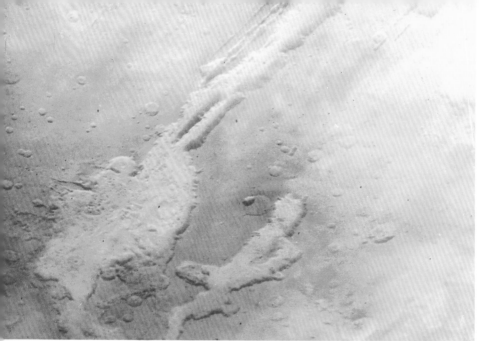

A vertical shot by a Viking orbiter of the giant Valles Marineris system, a system of gorges which dwarf the Grand Canyon in the South West of the United States (NASA/JPL).

Rocks of various sizes can be seen in this view of the Utopian Plain on Mars, taken by the camera on board the Viking 2 lander.

down. The task of the rover would be to study soil characteristics at several sites, with a drilling device capable of penetrating to several metres.

The 1996 Soviet mission will be concentrating on a sample return. This is even more complex than the 1994 flight and will involve the landing of a larger automated rover on the surface. This rover will have a radius of operation of up to 150 km from the landing site. Its operations will have to be based on artificial intelligence concepts, for it will be programmed to obtain and document samples and return them to the landing site. There, just as in the proposed American sample return mission discussed below, the samples would be transferred to a small rocket capable of returning them to Earth.

Luna 16 was a sample return spacecraft which obtained a small quantity of material from the Moon's surface in 1970. In the 1990s a rather more sophisticated robot craft will be sent to Mars on a similar mission (Novosti).

Sagdeev wrote the following in an article celebrating international cooperation in space (1988 *Physics Today* **41** May 30–8).

The likelihood of implementing the mission to Mars in 1994 and the sample-return mission before 2000 depends mainly on the development of space technology. This program requires a launch vehicle with a powerful upper stage, an interplanetary complex to be assembled in Earth orbit, and a spacecraft designed to assure aerodynamic braking in the Martian atmosphere while shifting from the cruise trajectory to Mars orbit. There are many other problems as well—in particular, the problem of the Martian soil quarantine on Earth. From our viewpoint, however, it looks like these problems can be solved in the years ahead.

He made it clear that international cooperation will be essential for both these missions. This is not so much financial, as it would be in the case of a US call for cooperation, but technological. Despite its brilliant successes the Soviet space establishment is not so well endowed with the very latest technology that it can afford to spurn assistance from other nations. A sample return mission involving both the US and the Soviet Union, and perhaps ESA and Japan as well, would be a logical result of the interdependence which now characterizes planetary missions.

In so far as NASA has a planetary exploration strategy—given the restraints which budgetary limits impose—it is based on two reports by the Solar System Exploration Committee of its Advisory Council. Although there is some opposition to this strategy on the grounds that resources might be diverted from other forms of scientific inquiry, the American scientific community is on the whole in favour of the idea of continued exploration of Mars.

In the first report, published in 1983, *Planetary Exploration Through Year 2000. A Core Program*, the committee recommended a number of missions in an ambitious programme. Firstly, the committee recommended the launch of a 'Geoscience Orbiter' to map the elemental and mineral composition of the martian surface on a global scale with a resolution of tens to hundreds of kilometres.

Secondly, they urged that a 'Climatology Orbiter' be placed into polar orbit around Mars for at least one martian year to record the expansion and contraction of the polar caps, to distinguish between water and carbon dioxide on the surface, to provide evidence of hydration and dehydration of surface minerals and to show how water vapour varies both regionally and seasonally.

They concluded that these two orbiters could be combined in one spacecraft in polar orbit around Mars, containing a gamma-ray spectrometer, a mapping reflectance spectrometer, a radar altimeter, a magnetometer, a thermal infrared radiometer/spectrometer and an ultraviolet spectrometer/photometer.

The committee's suggestion for a combined orbiter to examine both geoscience and climatology was accepted and a new programme, entitled Mars Observer, was commenced in the 1984 fiscal year.

This spacecraft was affected like all NASA programmes by the Challenger disaster and it will now be launched from the shuttle Discovery in September 1992, or two launch windows after the intended launch.

Like so many other US planetary spacecraft, Mars Observer has come from the Jet Propulsion Laboratory. It is a low-cost craft, based on existing technology and designs, and similar craft will later be used to study the Moon, the asteroids and Venus.

Mars Observer is in fact based on the design of the Satcom K communications satellite produced by RCA. It uses proven electronic sub-systems from the TIROS and DMSP series of meteorological satellites. It will be propelled towards Mars by an upper stage called the Transfer Orbit Stage (TOS) and after a one-year cruise phase will go into orbit round the planet. The high inclination orbit will be sun synchronous, so that the cameras will be able to map the martian surface thoroughly.

Its scientific tasks will be very much in line with those laid down by SSEC and so will the instruments it will carry. A gamma-ray spectrometer will measure the abundance of elements over the entire surface of Mars. A magnetometer will determine the nature of the magnetic field, and a camera will provide low-resolution global images and high-resolution images of selected areas.

Atmospheric temperature, water vapour and dust content will be measured by a pressure modulator infrared radiometer as they change with latitude, longitude and season, and a radar altimeter will determine the topography. A radioscience experiment will use the spacecraft's radio transmitter to map Mars's gravitational field and also to measure the atmospheric temperature structure. The mineral composition of the rocks will be mapped by a thermal emission spectrometer, which will also map frosts and the composition of clouds.

Finally, a spectrometer in the visual and infrared spectrum will provide a mineralogical map of Mars and chart concentrations of water and carbon dioxide. The original cost target was $212 million, but it is expected that this will be exceeded by some tens of millions of dollars. Even so, this will be a remarkably cheap mission compared with Viking, which cost just under $1 billion in 1977 dollars—nearer $2 billion in current values.

As the next stage in the exploration the SSEC recommended that an 'Aeronomy Orbiter' should be sent to Mars to study the magnetic field, how the planet interacts with the solar wind and the structure and dynamics of the upper atmosphere. They also urged NASA to mount a mission in which a simple orbiter would relay back to Earth data obtained by 'penetrators', probes which would be fired into volcanic areas of the martian surface like darts.

The primary goal of the penetrators would be to measure the bulk chemical composition of the rocks, including trace elements.

In 1986 the Solar System Exploration Committee published their second report, *Planetary Exploration Through Year 2000. An Augmented Program*, in which the rationale for continuing to explore Mars was more comprehensively spelled out. It said

> Whether or not Mars has ever been an oasis for life, the nature of the planet—and its direct relevance to understanding Earth and other planets—makes it a compelling target for in-depth exploration.
>
> A range of geologic processes has operated on Mars to produce landscapes that are alien and yet familiar; individual shield volcanoes that would stretch from Boston to Washington DC; a canyon that would extend from New York to Los Angeles; and seas of sand dunes that girdle the entire north polar region. Mars is thus a perfect natural laboratory for studying the various geologic forces that shape a planet.
>
> Mars is also a perfect laboratory to investigate planetary weather and climate. The diurnal and seasonal cycles on Mars are remarkably similar to those of Earth, but the extremely thin atmosphere, the rapid heating and cooling of the surface, the lack of oceans, and the enormous vertical scale of the landscape also create critical differences.
>
> The condensation cycle of atmospheric carbon dioxide and the exaggerated role of dust in heating the martian atmosphere are also fascinating contrasts to Earth. Despite these differences, we have already learned that martian meteorology is exciting, complicated, and capable of being understood, and a better understanding of the weather of Mars will teach us much about the weather of our own world.
>
> Martian climatology, which is in essence the long-term history of the planet's weather, is a science still in its infancy, but it holds great promise to illuminate not only the history of Mars but also that of Earth. There is ample evidence of massive climate changes on both planets—ice ages, changing seashore-lines, and species extinctions on Earth; regional flooding, glaciation and periodic polar sedimentary layering on Mars. Common mechanisms could have been at work on both planets—solar luminosity changes, periodic orbital variations, episodes of volcanic eruption, and asteroidal impacts. With detailed comparative data from both Mars and Earth, we can make important progress towards understanding matters of both intellectual and practical importance.

The SSEC particularly recommended a Mars Sample Return Mission 'because of its high scientific importance and because of its pivotal role in the future exploration of Mars'. On the scientific objectives for Mars exploration, the report said that NASA should attempt to do the following.

1. Characterize the internal structure, dynamics and physical state of the planet and the chemical composition, mineral composition and physical character of surface materials.
2. Determine the chemical composition, mineral composition and absolute ages of rocks and soil for the principal geologic provinces and the chemical composition, distribution and transport of volatile compounds (e.g. water, carbon dioxide) in order to understand the formation and chemical evolution of the atmosphere and the interaction of the atmosphere with the surface material (regolith).
3. Determine the quantity of polar ice and estimate the quantity of permafrost present on Mars and the processes that have produced the landforms on the planet.
4. Characterize the dynamics of the martian atmosphere on a global scale and the planetary magnetic field and its interactions with the upper atmosphere, incoming solar radiation and the solar wind.
5. Determine what organic, chemical and biological evolution has occurred on Mars and explain how the history of the planet constrains these evolutionary processes.

These global and regional data are needed for comparative planetology and to provide the spatial context for subsequent more sophisticated *in situ* measurements on the martian surface or within the atmosphere.

A network of surface stations could provide seismic data which could determine the internal structure, and also long-term meteorological data.

Although these missions can explore Mars and its environment to some extent, the SSEC report stated that by far the most accurate way to make many measurements was to return samples from the planet for detailed study and analysis on Earth. For this reason a series of Mars Sample Return Missions was recommended 'at the earliest possible opportunity'. Thirty pages of the report are devoted to a detailed rationale for this type of mission and also a description of the mission as it might be carried out in 1996 to 1998.

According to this schedule, the spacecraft would leave Earth orbit on 18 November 1996, and arrive at Mars 303 days later on 17 September

1997. After 401 days on the surface it would leave on 23 October 1998 and arrive back in Earth orbit on 14 September 1999, after a mission lasting 1030 days. At the start of the mission, after a Centaur upper stage has been used to propel the 'Interplanetary Vehicle System', or IVS, out of Earth orbit, the total mass of the system would be 9493 kg. After shedding various components the mass would be 8490 kg on arrival at Mars, where the craft would become the MOV, or Mars Orbiting Vehicle.

Orbit would be achieved by aerocapture, described in Chapter 5, and the MOV would separate into two sections, the Orbiter and the Mars Entry Capsule, or MEC. The lander would be de-orbited by means of aerobraking, also described in Chapter 5, and a parachute would be deployed in the atmosphere to slow the descent. Finally, a terminal descent engine would be used to complete the landing over the last few metres.

On landing, the capsule would deploy a 400 kg automated rover, driven by electric motors and capable of carrying out some geological

At a chosen site on the edge of a sinuous channel on the surface of Mars, a future sample return spacecraft lands to take tiny quantities of several different types of soil. NASA is actively researching ways of mounting this mission in the late 1990s (NASA).

tests of its own. Thus the results of simple tests of rocks, familiar to field geologists, such as with a Geiger counter or an acid bottle, could be reported back to Earth, so the mission scientists could decide whether samples should be taken at given sites. The rover would be able to take surface samples and also drill for cores.

The SSEC concluded that a sample collection strategy that maximized the number of samples weighing 0.5 to 10 g would also maximize the information return by providing the widest possible range of materials, the largest number of locations and the greatest opportunities for comparative study. A total of only 5 kg of samples would be collected during the 401 day surface stay and the rover would then return to the entry capsule and place the samples in a canister within a part of the lander called the Mars Rendezvous Vehicle.

The MRV would then be launched from the martian surface to rendezvous with the Orbiter, which would have been relaying telemetry back to Earth. The canister would be transferred to the Orbiter in a

An artist's impression of the sample return mission which has been studied by the Jet Propulsion Laboratory in California for NASA. The ascent stage is lifting off from the surface of Mars with a 5 kg sample of rocks and will rendezvous with an orbiter which will propel it back to Earth (NASA/JPL).

module known as the Earth Return Vehicle, which would fire its integral engines to place itself on a trajectory back to Earth. Finally, a 65 kg Earth Return Capsule would be separated from the ERV and would be manoeuvred into a 12 h Earth orbit by means of aerocapture, for retrieval either by the Space Shuttle or by an unmanned Orbiting Transfer Vehicle. The 20 kg Sample Canister Assembly would be extracted either on Earth or on board the Space Station which should be in orbit by then. The SSEC points out that the Space Station could play an important role by making possible preliminary analyses before they are taken down to the Earth, so limiting the risk of contamination of the terrestrial environment.

A sensible compromise may be to divide the areas of responsibilities, so that, for instance, the Americans might build the lander and the rover and the Russians the Orbiter and the Earth Return Vehicle. This method of international cooperation would avoid the awkward problems of financing joint craft and integrating one nation's launch vehicles with the other's spacecraft.

5 The First Explorers

The exploration of space is one of those vast inexorable movements of the human race, like the westward expansion of the United States. It is our manifest destiny.
Michael H Carr, US Geological Survey

When astronaut Sally Ride was asked by the administrator of NASA to report on the future of the agency, one of her recommendations was that America should send manned craft to Mars. In particular she advocated three manned missions with the aim of establishing an outpost on Mars by 2010.

This was merely the latest proposal for such a mission, for much thought and detailed analysis have been applied to planning manned missions to Mars by American specialists in the last three or four decades. No doubt similar calculations have been proceeding in the Soviet Union as well. It is now possible to visualize in broad outline how such a mission might be carried out.

Flying to another planet such as Mars is not a simple matter of pointing the spacecraft in the general direction of the planet and firing rockets. As mentioned in earlier chapters, launch windows for Mars occur every two years or so, but just as important to the design of the mission is the present state of technology. Rather than being equipped with rocket motors able to fire continuously, all spacecraft to date have been based on chemical rockets which fire for brief periods, imparting a velocity change, or delta V (Δv) to the spacecraft, to change altitude or orbit inclination or to place it in or remove it from an orbit around a target body.

Some mission profiles have assumed that nuclear or electric propulsion of some sort will have been developed by the time mankind wishes to go to Mars. The various types of electric propulsion which have been proposed operate by firing a rocket with a very small thrust but a high specific impulse over a long period of time, so building up a high velocity by small increments.

A quarter of a century ago this model of a Mars ship employing electric propulsion was produced by NASA's Lewis Research Center. A nuclear reactor at the front would be used to generate 30 Mwe and the electrical power would accelerate ionized gas to enormous exhaust velocities in the rockets. It may be another 25 years before anything like this ship flies (NASA).

This method would in due course make journeys to Mars extremely swift. In the extreme case, if it was possible to fire a rocket continually with a thrust that imparted an acceleration equivalent to 1 g to the spacecraft it would travel from the Earth to Mars in two days! This is not a feasible proposition in any foreseeable future. Even to develop electric propulsion with a small thrust at the scale which would be need for large manned spacecraft would probably take decades and it is not realistic to expect electric propulsion to be used in the first phases of Mars exploration. NASA did go some way in the 1960s towards developing NERVA, a nuclear rocket which was intended mainly for Mars exploration. The project was cancelled before a prototype rocket was ready and in the present climate of anti-nuclear feeling in the United States it is unlikely to be revived.

The various schemes which have been produced, on paper, to send men to Mars have often been distinguished by their impracticability. Werner von Braun, the V2 rocket scientist whose contribution to Project Apollo is incalculable, came up with the first hardware proposals as long ago as 1952.

In his *Das Marsprojekt* he proposed a mission on a truly heroic scale, with ten spacecraft each of 3600 tonnes and carrying seven men. Three of the ships would carry 200 tonne landing 'boats' each capable of landing 50 people on the surface for a 400 day stay. The fleet would be assembled in low Earth orbit with the aid of no fewer than 50 shuttles, each weighing more than 11 000 tonnes.

By 1956, when *The Exploration of Mars* was published in English, this mammoth project had been scaled down to a more reasonable two ships, each weighing a mere 160 tonnes. But even this would require 400 shuttle launches—two a day for seven months.

Another of the Peenemunde scientists, Ernst Stuhlinger, who had followed von Braun to the Marshall Space Flight Center, proposed a five ship mission in 1959. The ships would weigh 360 tonnes each. In a later version of this mission Stuhlinger proposed the use of a nuclear reactor to power an ion drive on a 450 tonne ship.

Later, both NASA engineers and aerospace contractors proposed more practical schemes based on the Apollo technology and using Saturn V and Saturn 1B rockets to assemble a Mars craft in low Earth orbit. North American Rockwell (now Rockwell International) proposed in 1967 a Mars Excursion Module of 30 tonnes. Geoffrey Canetti, the engineer responsible for this design, faced up to the question of how such a craft would survive entry into the martian atmosphere by proposing a heatshield with a diameter of 9 m.

The chance of using Apollo technology disappeared with the decision to discontinue the production of the Saturn rockets, and also the cancellation of the NERVA family of nuclear rockets, which could have been used with the Saturn as a powerful propulsive stage for a Mars mission.

Limitation to chemical rockets means that the trajectory to Mars must be carefully planned and accurate, since most of the energy applied to the spacecraft will come from the initial launch into Earth orbit and a single brief firing of an 'escape stage', together with minor mid-course corrections. The launch by a multi-stage rocket into Earth orbit will be achieved by reaching a velocity of $8 \, \text{km} \, \text{s}^{-1}$, sufficient to place the spacecraft in an orbit at an altitude of about 300 km.

The spacecraft may be launched in one large combination, or in the form of individual modules launched independently and combined in

The nearest NASA ever got to building an atomic rocket was this mock-up of the NERVA engine. The project was cancelled in the 1960s, but some experts believe that it should be revived to make a manned mission to Mars possible. It could even form the basis of a rocket in which energy was generated by the mutual destruction of matter and antimatter (NASA).

Werner von Braun, the Peenemunde engineer who joined the US Army rocket programme after the Second World War and then went on to become one of the prime architects of the Apollo Moon landing programme. He was one of the first to devise a mission concept for Mars exploration (NASA).

orbit, which will almost certainly be the preferred option. Once in orbit the spacecraft will still be accompanying the Earth in its heliocentric orbit at a velocity or $29.8\,\mathrm{km\,s^{-1}}$. A comparatively small velocity increment or Δv is then needed to throw the spacecraft out of Earth orbit and into its interplanetary trajectory.

In the case of Mars the extra velocity needed is of the order of $3\,\mathrm{km\,s^{-1}}$. In most cases studied this Δv will be achieved by means of a rocket stage which will be attached to the manned spacecraft during in-orbit assembly and may remain attached throughout the mission to act as a radiation shield in the event of a high-energy solar flare.

A second propulsive stage will be used to return to Earth orbit those components of the manned craft which are not discarded on the martian surface or in martian orbit.

In general, the only practical way to reach Mars with present technology is the method employed in all unmanned missions to date. This is by means of a cotangential transfer orbit. The spacecraft is placed in a heliocentric orbit originating at the Earth which at or near its aphelion intersects the orbit of Mars. Without a reduction in velocity the spacecraft would merely fly past Mars and remain in its heliocentric orbit, like the Mariner 4 craft.

But in the manned missions the rockets will be fired again to reduce the speed by an amount sufficient to allow Mars to 'capture' the spacecraft into an orbit around the planet, although Sally Ride proposes a different solution (see below). There is nothing particularly difficult about this process. The technique was used to place Apollo craft into lunar orbit and Mariner 9 and Vikings 1 and 2 into martian orbit. The Soviet Union have also used it to place Venera capsules into orbit around Venus and 'Mars' craft round Mars.

An exception to the general rule is the so-called Venus swing-by manoeuvre, in which the gravitational field of Venus is used to accelerate the spacecraft after its initial insertion into an orbit passing close to that planet, or to decelerate it during the journey back to Earth. However, this has the disadvantage of increasing the mission duration.

An added complication with a manned mission is that at least part of the spacecraft will have to be brought back to Earth. This is affected by another fundamental restraint. Just as the Earth and Mars have to be in a favourable geometry relative to one another at the time the craft is first launched, so the Earth must be in a suitable place if a cotangential transfer orbit from Mars is to reach it.

Missions which are of the opposition type, when all the activities take place just before and after the period when Mars is closest to the Earth, can include only a brief stay on Mars. On the other hand, the conjunction mission, in which the outward journey is conducted at one opposition, and the return journey at the next, can be much longer, with the advantage that the energy requirements are less.

Venus may be the first planet to be visited by men on their way to Mars in the early years of the 21st century. The Mars-bound craft would gain velocity by swinging close to Venus in a manoeuvre known as a gravity assist. This picture of the clouds that surround Venus was taken by Mariner 10 in 1974 with an ultraviolet filter which reveals patterns that are not seen in visible light (NASA).

If minimum energy is used in a conjunction mission—which maximizes payload for any given launch opportunity—the transfer time from Earth to Mars works out at 0.709 years (258.7 days).

The crew would then have to wait on Mars or in orbit round the planet for 1.278 years (466.5 days) before firing their motors and entering a trans-Earth trajectory. They would reach the vicinity of the

Earth in another 258.7 days, *so the total mission time would be 2.66 years (970 days)*†.

The Ride proposals described below advocate the shorter opposition type, and other opposition missions which have been devised employ Venus swingby to reduce mass in LEO (low Earth orbit). But the minimum energy mission would last three years, taking into account the preparation time in Earth orbit, although the flight crew would presumably minimize their exposure to weightlessness by joining the Mars craft only days before launch from Earth orbit, the assembly having been completed by a different crew.

Almost three years in space presents a daunting prospect, and the health implications are discussed in a later chapter.

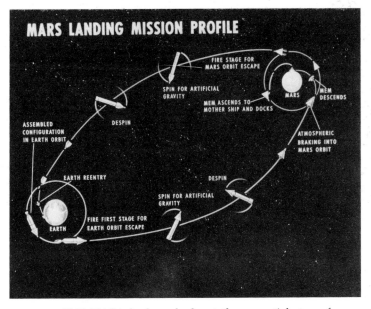

As long ago as 1964 NASA had worked out the essential steps for a manned mission to Mars. They remain much the same today, although some refinements such as a Venus swing-by and aerobraking on the return to Earth would reduce the mass of the spacecraft by a large factor (NASA).

†Roy A E 1988 *Orbital Motion* (Bristol: Adam Hilger) 3rd edn

What type of spacecraft will be needed to make the first Mars manned journey? In the first place, the component parts of the spacecraft will be placed in Earth orbit by means of Shuttles and their derivatives and a newly developed Heavy Lift Launch Vehicle (HLLV) if the mission is American, and by the Russian Shuttle and Energiya rocket if it is Soviet.

The craft will have to include a large rocket to propel the whole complex on the transfer orbit towards Mars, a crew transfer module, or 'mother ship', in which the crew will live during the two legs of the journey, a descent craft, or 'Mars Excursion Module' capable of surviving the entry into the martian atmosphere and landing softly in a chosen location, and an ascent craft to take the landing party back to a rendezvous with the orbiting mother ship.

Finally, a propulsion stage to return the crew transfer module from martian orbit to an Earth-bound trajectory and a means of slowing the crew transfer module down from its interplanetary velocity to a speed at which it will be captured into Earth orbit will be required. As discussed in Chapter 6, this will probably be a combination of retro-rocket and aerocapture.

The technique of martian orbit rendezvous will be used to save weight, just as lunar orbit rendezvous was used in Apollo. The propellant which is needed to return the crew to Earth orbit is never landed on Mars but remains in orbit in the mother ship, since it would need vast quantities of energy, and therefore propellant, to land it on the planet and return it to orbit.

This equation may change radically in future missions, for it is feasible to manufacture rocket propellant on the surface of the planet. One relatively simple way would be to extract water from the planet's crust and reduce it to its constituent oxygen and hydrogen. Since most of the engines used on Mars missions will be propelled by these two elements, just as the Space Shuttle main engines are, a priority for a martian base could be to construct just such a plant. Merely heating the soil to a sufficiently high temperature is sufficient to drive some water out of it, so the plant need not be unnecessarily complicated.

From the results of the Viking lander studies it is known that 100 kg of martian soil contains 1 kg of water, so it is not a rare resource.

Another scheme which has been studied by NASA is to obtain oxygen from the martian carbon dioxide atmosphere. In a tentative scheme the CO_2 is thermally dissociated into carbon monoxide and oxygen which are then separated by a solid electrolyte membrane. The oxygen is available as an oxidizer, with methane as the fuel, but only the methane then needs to be taken to the martian surface. NASA is seriously considering this system for unmanned missions in the future, but it would be even more useful in manned flights to Mars.

In the early 1960s NASA allowed its artist's imagination to wander in this impression of a four-ship expedition preparing to leave Earth orbit for Mars. The project would have depended on the construction of the gigantic Nova rocket, with a payload of 450 tonnes (NASA).

Something like the RTG (Radioisotope Thermoelectric Generator) used so successfully to power the two Voyager spacecraft over more than ten years in their journeys through the outer Solar System might be adequate for this task. The Voyager RTGS each had an output of only 160 watts but there should be no major problem in scaling up a device of this type, which depends on the radioactive decay of plutonium-238 to provide energy which is converted into electricity. Fuel cells and wind power are among other options that are being investigated by NASA.

Missions to Mars, it must be faced, need to surmount a number of technological hurdles before they can be mounted. But the scheme proposed by Sally Ride avoids some of the pitfalls of a very long mission by using some ingenious stratagems.

Although it remains true that a conjunction type mission to Mars must take about 1000 days, and even an opposition mission lasts two years, the Ride scenario visualizes three 'sprint' missions each taking only about a year.

In her report to the NASA administrator, *Leadership and America's Future in Space*, Ride says

> This leadership initiative declares America's intention to continue exploring Mars, and to do so not only with spacecraft and rovers, but also with humans.
>
> It would clearly rekindle the national pride and prestige enjoyed by the US during the Apollo era. Humans to Mars would be a great national adventure; as such, it would require a concentrated massive national commitment—a commitment to a goal and its supporting science, technology and infrastructure for many decades.

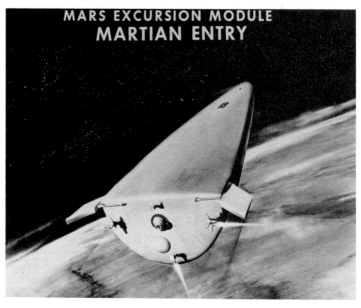

The 1964 concept included a Mars Excursion Module with a mass of 28 tonnes. It was of the lifting body type with a lift–drag ratio designed to take advantage of the tenuous martian atmosphere to make a precision landing possible (NASA).

There are three stages in the strategy which Ride proposed America should employ.

1. Comprehensive robotic exploration of Mars in the 1990s.
2. An aggressive space station life sciences research programme to validate the feasibility of long-duration spaceflight. This would develop an understanding of the physiological effects of long-duration flights, of measures to counter those effects, and of medical techniques and equipment for use on the flights.
3. The design, preparation and execution of three fast piloted round-trip missions to Mars, enabling a commitment to be made to establish an outpost on Mars by 2010.

The apparently impossible goal of cutting a three-year journey to twelve months would be achieved by sending a cargo vehicle ahead of the manned craft on a slow, low-energy trip to the planet, so minimizing its propellant requirements and reducing the LEO mass. Both this 'freighter' and the manned craft would be assembled in LEO but the men, or men and women, who would make up the crew would not leave LEO until the freighter had arrived in Mars orbit.

The six astronauts on board would rendezvous in Mars orbit with the freighter and refuel their craft from the waiting propellants. They would descend to the surface in a 'Mars Excursion Module', stay there for 10 to 20 days, and then return to martian orbit to rendezvous with the craft in which they had travelled from Earth. As this had been refuelled with propellant from the freighter it would then be able to return by a fast route to LEO, where the crew would transfer to the orbiting Space Station.

Ride says

> Even with separate cargo and personnel vehicles, and technological advances such as aerobraking, each of these sprint missions requires that approximately 1100 tonnes be lifted to low Earth orbit. In comparison, the Phase 1 Space Station is projected to weigh approximately 210 tonnes.
>
> It is clear that a robust, efficient transportation system, including a heavy-lift launch vehicle, is required. The complement of launch vehicles must be able to lift the cargo and personnel required by the sprint missions to the Space Station in a reasonable period of time.
>
> The Phase 1 Space Station is a crucial part of this initiative. In the 1990s, it must support the critical life sciences research and medical technique development. It will also be the technology test bed for life-support systems, automation and robotics, and expert systems.

This 1964 concept of a NASA mission to Mars owes more to science fiction than to the realities of space technology of the time. However, the picture shows remarkable prescience in depicting a deep chasm at a time when the canyons of Mars were still unrevealed (NASA).

Furthermore, we must develop facilities in low Earth orbit to store large quantities of propellant and to assemble large vehicles. The Space Station would have to evolve in a way that would meet these needs.

This initiative would send representatives of our planet to Mars during the first decade of the 21st century. These emissaries would begin a phase of human exploration and reconnaissance that would eventually lead to the establishment of a permanent human presence on another world.

A successful Mars initiative would recapture the high ground of world space leadership and would provide an exciting focus for creativity, motivation and pride of the American people. The challenge is compelling and it is enormous.

The 'sprint' missions would only become possible with progress in a number of new technologies.

Apart from the necessity to store cryogenic propellants for long periods in space, and transfer them from one craft to another, the sprint mission will demand advances in automation, robotics and interplanetary propulsion. Highly efficient rocket engines with a higher specific impulse—each increase in this index reduces the mass which has to be lifted off the Earth's surface—will have to be designed and built, and the control of the whole craft will largely be a matter of artificial intelligence. The complex will be so advanced and will demand the control of so many parameters that only some form of AI will be sufficient to keep everything in check, even though the crew will have manual control as a last resort.

The heavy-lift launch vehicle (HLLV) will also have to be developed. The Ride report does not specify exactly how many of these big rockets would be needed for each of the sprint missions, but the sketches used to illustrate it imply that there would have to be a total of six launches per mission. Assuming that the payloads would be equal in mass, this implies an HLLV payload approaching 200 tonnes. This is close to the upper limit for an HLLV based on Space Shuttle propulsive units, but it is not impossible.

An outline sketch of a possible HLLV has been published by E Brian Pritchard and Robert N Murray of the Space Station Office at NASA's Langley Research Center in Virginia, based on a design from Marshall Space Flight Center.

It would be, they say, in the 200 000 lb payload class, or approximately 90 tonnes. It would consist of a liquid-fuelled core with three of the Space Shuttle main engines, plus two solid rocket boosters of the type currently used on the Shuttle. The payload would be contained within a cylindrical jettisonable payload fairing 11 m in diameter and 36.50 m long. The overall weight would be 2060 tonnes and the payload capability would be 67 to 90 tonnes in a 400 km orbit at 30° inclination to the equator.

With this class of launch vehicle, which seems distinctly unambitious by comparison with the Soviet Energiya, discussed below, the three sprint missions proposed by Ride would demand a huge number of launches. Pritchard and Murray estimate that the three sprints would demand an HLLV launch every 45 to 60 days for eight years, a total of 50 launches of this complex vehicle. This may sound a tall order, but it should be remembered that NASA, in a period much closer to the beginning of the space age, were able to launch 13 of the bigger Saturn V rockets between April 1968 and May 1973, or one launch every 150 days. The added complication in the case of the Saturn V was that it was 'man-rated', with all the necessary extra care that has to be taken to protect the crew. The HLLV launch campaign would be

1970 - 1990

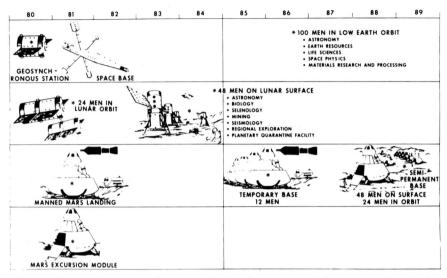

| 80 | 81 | 82 | 83 | 84 | 85 | 86 | 87 | 88 | 89 |

This is how the Marshall Space Flight Center in Alabama visualized the progress of Solar System exploration in 1969. If their plans had come to fruition there would now be 48 men on the surface of Mars—women were not included in NASA plans twenty years ago (NASA/MSFC).

easier because manned trips would be made in the Space Shuttle.

The thirteen Saturn V rockets orbited about 1200 tonnes, roughly comparable with the mass of one of Ride's sprint missions. In any case, as demonstrated in the next chapter, there are other versions of HLLV on the drawing board which would accomplish the task with fewer launches.

Apart from the construction of the HLLV, perhaps the biggest leap forward necessary for the Ride missions will have to be in aerocapture. In every case where an unmanned spacecraft has been put into orbit round the Moon, Venus or Mars, the necessary reduction in the spacecraft's velocity has been achieved by rocket propulsion.

A 'retrorocket' is fired in a direction contrary to that in which the craft is travelling to reduce the speed of approach to that which is needed to allow it to fall into orbit. In the Ride plan this would not be necessary. Instead, both the freighter and the manned transport would be equipped with highly efficient heatshields like those used successfully on the Command Module of the Apollo on its return from the Moon. The trajectory of each craft would dip into the martian atmosphere just deeply enough to bring about the necessary change of velocity. The craft would then 'skip' out again into an elliptical orbit of which the lowest point, or periapsis, would still be within the atmosphere. At the highest point, or apoapsis, a small rocket burn would be sufficient to raise the periapsis above the atmosphere.

It is claimed that a reduction of 50 per cent in the mass of the vehicle can be achieved by this technique.

Advances in adaptive guidance technology, thanks to the development of faster and more capable microprocessors, make aerocapture a much better proposition than it would have been during the days of fairly crude inertial guidance and navigation systems.

However, even a decade seems a short time to develop heatshields on the scale contemplated by the Ride report. The Apollo command module heatshield, the largest so far flown, had a diameter of 3.9 m; the freighter which Sally Ride advocates and the manned transfer craft, both of which would depend on aerocapture to go into orbit round Mars, would each need a faultless heatshield of more than double that diameter and perhaps as much as 36 m.

This figure arises from research done into aerocapture at NASA's Ames Research Center in California. There, a basic design for an aerocapture vehicle for return from the Moon to the Earth has been devised. It consists of a 'spherically blunted, raked-off sphere–cone' and in the lunar form it would have a diameter of 12 m.

In a paper to the IAF Congress in 1987 Gene P Menees, of Ames, said that a similar aerocapture vehicle would be suitable for both entry into martian orbit and descent to the surface, but the diameter would have to be two or three times as great.

A similar cone–sphere of only 12 m diameter would suffice for the Earth-return portion of the mission. Menees was able to show that if retropropulsive means only were used to place a returning Mars capsule into LEO the required Δv would be almost 6 km per second. In contrast, if aerocapture was used only a trivial amount of propellant would need to be burned to circularize the original elliptical orbit.

If retropropulsion were used the propellant needed would take up 70 per cent of the vehicle mass and would probably make the mission impossible. Menees hypothesized a return capsule of 5 tonnes with a heatshield, or more accurately 'aerobrake', diameter of 12 m. This would be the last remaining module of a manned craft of 134 tonnes to go into orbit around Mars. These figures do not match up with the

Ride estimates, but that does not mean that the Ames research will not be relevant to an eventual manned mission. Ride's ideas were essentially painted with a broad brush and practical considerations such as the Ames research will refine them in due course.

Some hardware research, in addition to theoretical studies, is going on within the NASA establishments into the concept of aerocapture and its application to bringing back satellites from geosynchronous orbit, but the research will presumably also have Mars applications.

In the Space Station era there will be some activity in the geosynchronous zone, 40000 km above the Earth's surface, and from time to time it will be necessary to use an orbiting transfer vehicle (OTV) or space tug to bring a satellite back to the Space Station for replenishment or repair.

In pursuance of this technology, the Johnson Space Center in Houston is building, at a cost of $179000000, an aerocapture vehicle with a heatshield diameter of 4.26 m. The 'Aeroassist flight experiment', or AFE, will be launched from a Space Shuttle in 1993 and will simulate the return of a space tug from geosynchronous orbit. A solid rocket motor will propel AFE from low Earth orbit to the fringes of the atmosphere at a height of around 76 km before soaring back up out of the atmosphere. In doing so it will reduce its excess velocity—the result of the rocket burn—and remain in Earth orbit.

Another study of aerobraking and aerocapture for Mars missions has been carried out by James French, of the Jet Propulsion Laboratory. Aerobraking is a technique which can be used when a spacecraft already in a highly elliptical orbit uses drag during successive passes through the upper atmosphere to circularize the orbit. French found that aerobraking could be used to place a space station or orbital fuel store into a circular martian orbit after a propulsive manoeuvre had placed it in an elliptical orbit.

A major advantage is that it can increase the mass of the spacecraft in final orbit by up to 70 per cent, and reduce the total Δv requirements by 1 to 1.5 km s^{-1}. These results are broadly in agreement with those reached by Menees, both as far as Δv requirements and mass saving are concerned.

French also found that aerocapture was a more effective technology than retropropulsion, especially for high approach velocities compatible with manned interplanetary flight. But instead of a spherically blunted, raked-off cone, favoured by Menees, he visualizes a biconic vehicle, similar in all aspects but size to the re-entry vehicles of nuclear missiles. Using a variable lift/drag ratio this vehicle would be steered into the atmosphere until it has reached the velocity necessary for capture by the planet into a low orbit. It would then 'skip' out of the atmosphere and a small burn of a rocket system at the required altitude would circularize the orbit outside the atmosphere.

When the decision is made to leave the low orbit and land on the surface the biconic vehicle is capable of great precision in landing. The crew would be able to steer it to a predetermined point—or to an established base—with braking rockets being used during the last few hundred metres of the descent. This would be of great importance if, for instance, supplies had been landed in advance of the first manned landing as a safety measure.

The biconic 'nose-cone' is still merely an outline drawing, but it has the potentiality of forming a combined descent and ascent vehicle, which lands vertically. The upper portion of the bicone is then employed as the ascent vehicle, just as the ascent stage of the Apollo Lunar Module lifted off using the descent stage as a launch pad.

The exploration of Mars begins. In this illustration of the 'Case for Mars' mission, three biconic Mars Modules have landed on the surface and astronauts have emerged to begin their studies of the planet. The upper section of the lander is the ascent stage, in which the crews will fly up to rendezvous with the passing interplanetary space station on its next approach to Mars (Carter Emmart).

A variety of mission profiles, depending on energy requirements, have been drawn up by American enthusiasts. From time to time the group of scientists, engineers and other enthusiasts who are known as the 'Mars Underground' hold conferences in Boulder, Colorado, entitled 'The Case for Mars'. In *The Case for Mars II*, in 1984, a biconic vehicle design was much discussed and some artist's drawings were later produced to illustrate how it might look, both as the tip of the trans-Mars vehicle and on the martian surface.

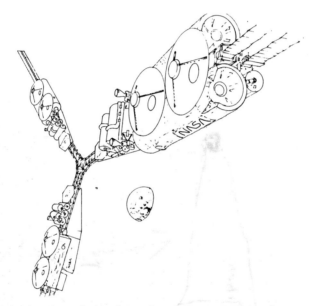

An artist's impression of the 'interplanetary space station' proposed in the 'Case for Mars' study. Each of the three arms bears a biconic Mars Module for landing astronauts on the martian surface. The station continues on a heliocentric orbit back to Earth and will return in due course to rendezvous with the Mars Modules at the end of their mission (Carter Emmart).

In a paper presented at the International Astronautical Federation's Congress in Brighton in 1987, Michael B Duke of the Johnson Space Center gave some comparative figures for different types of missions. The 'sprint' mission proposed by Sally Ride involved placing 1200 tonnes into low Earth orbit and 25 tonnes onto the surface of

Mars. The mission duration would be 14 months and the stay time on the surface 10 to 20 days. An opposition class mission would need 700 tonnes in LEO, would be able to place 65 tonnes on the martian surface, and would have a mission duration of 21 months, with 2 months spent on the surface.

Finally, a conjunction class mission would need only 600 tonnes in LEO and could land 100 tonnes on the surface for a 15 month stay. This would be the minimum energy case, but the snag would be the 35 month mission duration.

The Ride report has been received with enthusiasm by NASA and even received some endorsement from President Reagan during his last year in office. On 11 February 1988, he spelled out a National Space Policy, which called for a commitment to 'manned flight beyond the Earth into the Solar System', but stopped short of actually endorsing a manned mission to Mars. In addition, the president backed a 'Pathfinder' programme aimed at improving America's space technology.

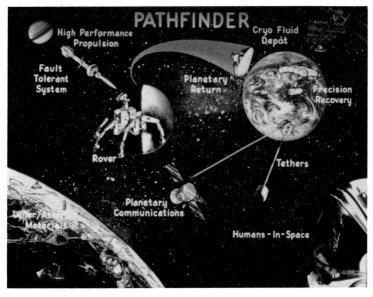

Project Pathfinder has been initiated by NASA to develop the technologies which will be necessary to continue the exploration and exploitation of the Solar System. One of the main objectives is the development of technology to support manned Mars missions (NASA).

Among the expenditure proposed was $14 million on orbital transfer vehicle (OTV) technology, vital in assembly of a Mars craft, $13 million on long-term manned flight issues, $41 million on deep space operations technology and $15 million on lunar and martian studies.

Contracts have already been awarded to aerospace companies for studies of mobility of a robotic rover on the surface of Mars, and aerobraking for landing a sample return mission on Mars.

Pathfinder will mature into a multi-year $1 billion programme with the avowed aim of restoring America's lead in space and giving NASA the ability to return astronauts to the Moon and begin flights to Mars early in the 21st century.

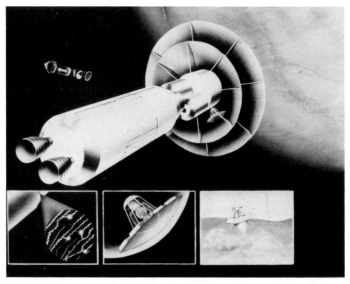

Two big aerobrakes—one for entering the atmosphere of Mars and the other for the return to Earth—dominate this picture of a projected spacecraft. It was produced by NASA to illustrate the sort of technology which will be developed under Project Pathfinder (NASA).

Under the programme a high priority has been given to the development of a Mars surface sample return mission in the late 1990s. An advanced 'rover' will be developed, capable of an autonomous landing on the surface of Mars and obstacle avoidance once it gets there.

Robotics will be developed to select, analyse and store samples before their return to Earth.

Automatic rendezvous and docking with a high level of reliability will be developed for both manned and unmanned missions, and orbital transfer vehicles will be designed, using liquid hydrogen/liquid oxygen (LH/LOX) technology. It has been estimated in this context that an increase of 35 seconds in specific impulse of the rocket motors employed could reduce the weight of a manned vehicle on the way to Mars by as much as 158 tonnes.

Pathfinder will also be concerned with power generation on the surface of the Moon and Mars and in flight, including nuclear power, optical communications, fault tolerant systems and the production of oxygen on the Moon.

Astronauts lay out a huge array of photovoltaic cells on the surface of Mars. Under Project Pathfinder NASA is developing the technology needed for high-performance, low-mass solar power arrays needed to support missions to the Moon as well as to Mars (NASA).

Several NASA establishments have now begun studying the possibility of manned missions to Mars. The Johnson Space Center in Houston will assess future lunar and Mars manned missions and, in particular, how a lunar base would be a stepping stone to Mars.

Goddard Space Flight Center near Washington will assess the scientific potential of such missions, and the Marshall Space Flight Center in Huntsville, Alabama, as well as laying down specifications for types of orbital transfer vehicles needed, is looking at propulsion requirements for both manned and unmanned operations between the Earth, the Moon and Mars.

Marshall, where much of the Apollo technology was born, has developed a concept in which the Mars transport would be split in two. Rotating the two halves at either end of a 60 m tether would generate a gravitational force of 1g, so reducing problems associated with weightlessness. This is described in more detail in the next chapter.

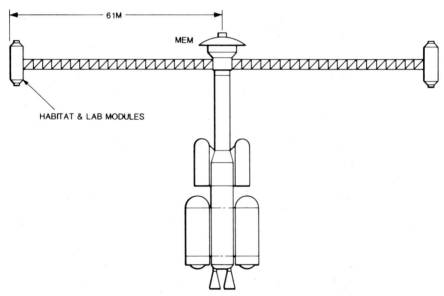

In this version of a spacecraft to take men to Mars, the MEM (Mars Excursion Module) is at the centre of a structure which rotates to give astronauts in the two habitat modules at the end of 61 m arms the equivalent of terrestrial gravity. Two big rocket stages provide the energy to get the craft to Mars and return it to Earth (NASA/MSFC).

Another facet of the Marshall studies is the possible use of the pressurized modules being produced by the Boeing company for the space station as crew quarters for a Mars expedition. The Langley Research Center in Virginia will study ways in which the space station itself could be used as a stepping stone to Mars.

Meanwhile, Soviet officials have stated quite bluntly that there is an intention to send men to Mars, but quite how it would be done has not been described.

Roald Sagdeyev, the leading civilian in a programme largely dominated by the military, has said the most about Soviet intentions. On visits to Washington he has openly invited the United States to take part in the venture, which he has suggested most recently would be mounted in the year 2007.

This is not a particularly favourable launch window. The mass which needs to be placed in LEO varies from launch window to launch window and in 2007 is about 1800 tonnes for a particular type of opposition mission, compared with about half that mass for 2001 and 2013. The first date is undoubtedly too early for a cooperative venture and 2013 seems much more likely.

While in Washington, Sagdeyev estimated that the cost of such a mission would be between $50 and 100 billion at present day prices, which is well within the range of the cost of some advanced military programmes, such as the B2 bomber mentioned in Chapter 8.

The deputy director of the Institute of Space Research, Vyacheslav Balebanov, told the Krasnaya Zvezda (Red Star) newspaper in December 1987 that the manned exploration of Mars remained the prime Soviet goal in space. Veteran cosmonaut Yuri Romanenko has also been quoted as saying that a manned mission to Mars was becoming a more and more realistic possibility.

These are only a few of the declarations of intent by Soviet officials on the subject of a manned Mars flight. They have undoubtedly been encouraged to be more optimistic by the successful first flight of the Energiya rocket.

Apart from being the largest rocket to be launched to date it was notable for being the first exhibited by the Soviet Union to have a second stage using the technology of liquid hydrogen/liquid oxygen propellants, employed by the Americans for more than twenty years. This has the advantage of a higher specific impulse, so that a rocket of a particular size employing LH/LOX can launch a larger payload than the same rocket propelled by hydrocarbon fuel plus LOX.

America, of course, has no rocket in this class, having decided to

83

A model of the huge Soviet launcher, the Energiya, shows three of the four strap-on rockets around the central core. The payload module is on the other side. This rocket is capable of placing a Mars-bound manned spacecraft into low Earth orbit with three or four launches (author).

cease production of the Saturn V after six successful Moon missions and the launch of the Skylab space station. Energiya, an ingenious design using 4, 6 or 8 liquid-propelled strap-on rockets round a central core, will eventually have a payload into LEO of 'more than 100 tonnes', in the words of more than one Soviet commentator. Evidently the exact payload weight is being kept under wraps until full development has been reached, but the Energiya will have three times the payload of the US Shuttle.

This will make it ideal for a manned Mars mission, for it should be possible to assemble a large spacecraft of 400 or 500 tonnes from modules launched by three, four or at the most five Energiya launches, while an exclusively Shuttle-launched mission would tie up the entire US Shuttle fleet for a long period, perhaps two or three years.

A much more likely solution for the Americans will be the develop-

Part of the Viking 2 lander is seen in this picture from Utopia Planitia, where the landscape was remarkably similar to Chryse Planitia. By choosing safe places in which to land the JPL scientists had placed the landers in relatively uninteresting terrain (NASA/JPL).

Viking 1 dug a trench with its soil sampler and placed some of the sample in the automatic biological 'lab', which provided some interesting results, but no unequivocal signs of life (NASA/JPL).

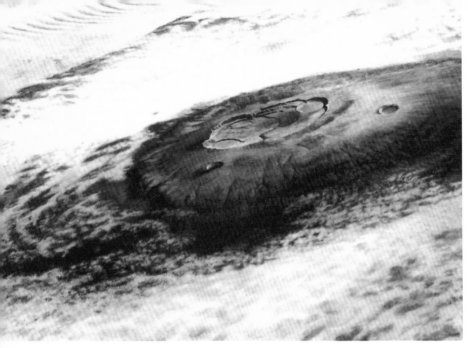

This striking Viking picture of Olympus Mons, the biggest volcano in the Solar System, shows the huge cliffs which have been formed by repeated lava flows. Clouds can be seen on the slopes of the volcano (NASA/JPL).

This view of the pink sky and rocky plain of the Utopia area of Mars was taken by a camera on board the Viking 2 lander. The horizon appears to be sloping because the lander came to a rest at an angle, with one of its footpads on a rock (NASA/JPL).

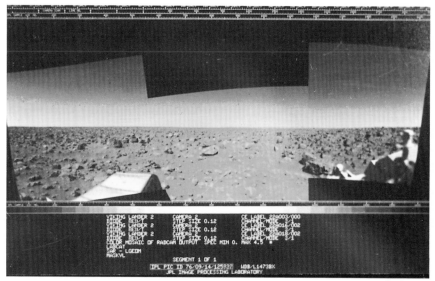

Pictures of the martian surface taken by the Viking landers included computer data on a screen on the craft. In future missions even more complete documentation will be needed (NASA/JPL).

The Soviet Union placed two Lunokhods on the Moon in the 1970s to carry out scientific research. A more sophisticated version is expected to go to Mars in the 1990s (author).

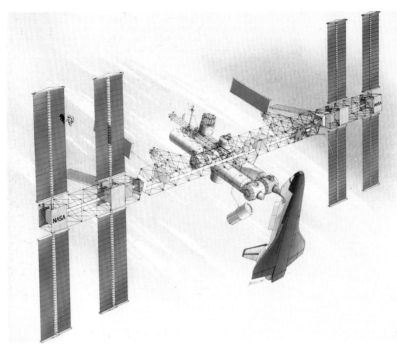

The NASA Space Station will be in operation from the middle 1990s. It may one day act as a base for the operations in low Earth orbit which will be necessary for the assembly of a manned spacecraft heading for Mars (NASA).

The pressurized modules in which astronauts will live and work when the US Space Station is operating in the 1990s might be adapted to serve as mission modules for a manned flight to Mars (NASA).

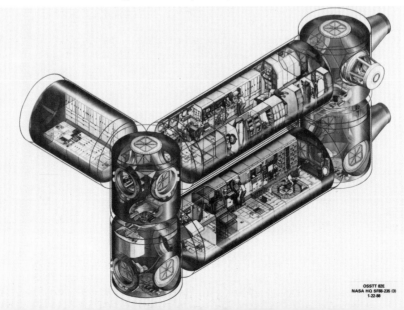

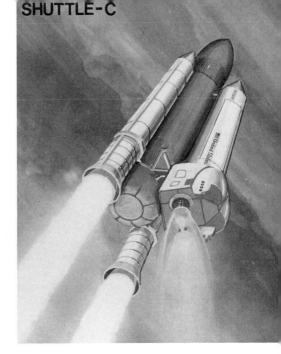

SHUTTLE-C

The Shuttle-C, an unmanned version of the Space Shuttle, is a concept of engineers at the Marshall Space Flight Center. It would have the external tank and solid rocket boosters of the Shuttle, but a large payload module in place of the Shuttle airframe (NASA/MSFC).

ment of a bigger version of the HLLV which was advocated in the Ride report. In this concept, elements from the Space Shuttle, such as improved solid or liquid boosters, the external tank for the liquid hydrogen and oxygen, and an advanced version of the Space Shuttle main engines (SSME) would be used to construct an enormous unmanned rocket. One concept, devised by Milton Page of NASA's Marshall Space Flight Center, would also use a new engine burning hydrocarbon fuel. It would be 120 m high and would be able to place a 185 tonne payload into LEO in a fairing measuring 13.7 m by 61 m.

6 The Big Boosters

In the long run, the binding up of the wounds on Earth and the exploration of Mars might go hand in hand, each activity aiding the other.
Carl Sagan

Page's description of the HLLV is to be found in a publication which can best be characterized as the bible of the sect which is urging the US Government to send men to Mars. The impressive two-volume work, *Manned Mars Missions* (NASA M001 and M002, June 1986) is the result of a conference at the Marshall Space Flight Center in Alabama in June 1985. A working group had been studying all aspects of the missions and presented more than 90 papers distilling the work of hundreds of scientists and engineers in NASA establishments, the Los Alamos National Laboratory, universities and aerospace contracting firms.

The level of ingenuity and technical expertise displayed by these advocates for men on Mars is high enough to ensure that the twin volumes will be on the bookshelves of every serious planner of such missions, wherever they work. Every conceivable problem has been studied in depth—mission concepts, spacecraft design, power supplies, aerobraking, launch vehicles, solar sails, budgets and cost, life sciences, the Space Station, even politics.

The editors of an overview of the conference made the following declaration.

> An opportunity to conduct the first manned visits and to establish the first settlement on another planet in the solar system is within the grasp of humans on Earth.
>
> Initiation of this historic enterprise will require a national commitment. It is believed that most Americans would find it unacceptable for political, economic, scientific or intellectual reasons for some other nation to dominate this effort. Politically, US leadership would insure a stable and open programme with a wide variety of international participants. Economically, the development of the technology and the space operations base necessary for these missions would bring new

By 1976 the scientists at the Jet Propulsion Laboratory in California could visualize a manned expedition to Mars using the advanced concept of solar sails. Few experts now see this as a possibility within the next 20 years (NASA/JPL).

commercial opportunities to Americans in a growing space-oriented enterprise.

Scientifically, leadership in planetary and space sciences brings benefit with better understanding of the planetary processes (atmospheres, oceans, tectonics, and the origin of life) on other planets as well as on Earth. Intellectually, 'going to Mars' is an endeavour that can captivate and motivate a generation of young people from throughout the world, just as the American frontier motivated generations past.

It must also be emphasized that the potential for international cooperation on a Mars venture is real and has far-reaching implications. Other nations with similar space capabilities and motivations also envision a prosperous future in space. Sharing the costs and benefits of a Mars programme with other nations in a project of this magnitude which has only peaceful objectives could also increase cooperation and decrease antagonism on other issues.

The 1000 pages of the reports are not easy to summarize. Some of them seem a little dated now, for they concentrate on plans for missions in the years from 1999 to 2001, but the budgetary difficulties of NASA ensure that there is now absolutely no prospect of any such missions being mounted. Nevertheless, much of the hardware proposed would be just as valid for missions at later launch opportunities.

Milton Page's paper, *Earth-to-orbit launch vehicles for manned Mars mission application*, is a fascinating insight into the rockets that will be needed once an expedition is planned. He points out that such missions will require payloads to LEO that are much heavier and larger than the Shuttle can accommodate. Assembling many small modules in LEO will make the missions more complicated and expensive.

Both NASA and the Department of Defense are studying new, larger rockets and Page described three of them. The first is the SDV-3R, discussed in the previous chapter. The first or booster stage would use the standard Shuttle solid rocket booster, the cause of the Challenger disaster in 1986, which has now been redesigned to make it safer.

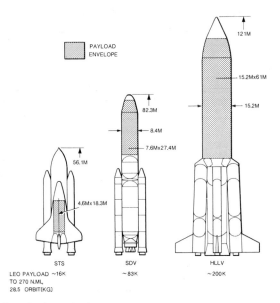

The three launch vehicles which advocates of Mars manned mission hope will support the venture. The Space Shuttle, with a payload of about 16 tonnes, is dwarfed both by the Shuttle Derived Vehicle (SDV) and the Heavy Lift Launch Vehicle (HLLV) (NASA).

The second stage of the SDV would be the external tank of the Shuttle. As flown in Shuttle missions this is a purely propellant tank, but in the SDV it would be fitted with a unit at the base including three of the Space Shuttle's main engines (SSME) and the avionics controlling the flight.

This 'P/A' module would be recoverable after being parachuted into the sea and the engines and avionics could be used again and again.

Above the core stage would be a payload shroud measuring 27 m × 8 m and capable of taking a payload of 86 tonnes to an orbit 300 km above the Earth. If the Centaur G Prime cryogenic upper stage was included as part of the payload it could boost a smaller payload to a higher energy orbit. Page optimistically allocates a five-year development period to the SDV-3R and points out that although some new facilities, such as a Mobile Launch Tower, would be needed the rocket could largely be operated from existing buildings and pads at Cape Canaveral.

Next, Page described a 'Shuttle derived heavy lift vehicle' to take payloads greater than the SDV-3R could handle. As an example of this concept he illustrates a rocket which has reusable strap-on boosters in place of the solids. These would be 43 m long with a diameter of 6 m and would each have two colossal engines burning LOX and hydrocarbon fuel. They would be advanced technology versions of the F-1 engines which powered the first stage of the Saturn V rocket.

The central core of the rocket would once again be an adapted external tank with three SSME engines and the payload shroud would have the same dimensions as the SDV-3R.

The difference would be that with the added thrust of the liquid boosters the rocket would have a payload of 136 tonnes to LEO, and Page estimates that the cost per kilogramme of payload would be lower.

The third vehicle he described is a heavy lift launch vehicle with a payload of 185 tonnes to LEO—a rocket that would be largely comparable with the Soviet Union's Energiya. New launch facilities and even launch sites would be needed for this gigantic rocket, 121 m high and with a payload shroud measuring 60 m × 15 m. New engines burning either LH/LOX or hydrocarbon fuel with LOX would have to be developed to give high performance and Page estimates the development time as ten to twelve years, so unless there is an early commitment to this project by the US administration it will be well into the next century before the rocket flies. Since the Energiya is already in its early operational life it may be launched dozens of times before the

HLLV appears. This is the real gap in the space performance of the two great superpowers, for US lead in such matters as micro-electronics and instrumentation is more than balanced by this disparity in propulsive capability.

The efficiency of a rocket engine is measured by an index known as specific impulse, or I_{sp}, which defines the thrust of the engine per unit rate of expansion of propellant weight. I_{sp} is expressed in seconds and is calculated by dividing the exhaust velocity of the rocket by gravitational acceleration at the Earth's surface.

The different specific impulses of the various types of rocket being proposed for a manned Mars mission demonstrate the relative efficiency of the types. The conventional type of chemical rocket burning liquid oxygen and liquid hydrogen, as in the Space Shuttle and most proposed Mars spacecraft, has a specific impulse of 468 seconds. In contrast, the type of engine which burns 'storable' propellants such as nitrogen tetroxide and monomethyl hydrazine and so does not suffer from the problem of boil-off of liquid hydrogen over long missions, has a specific impulse of about 346 seconds. The result is that very much more propellant has to be carried for this type of mission.

The attraction of more exotic types of engine is that they have a higher I_{sp} and so need less propellant. The NERVA rocket, using a nuclear reactor to heat hydrogen to a high exhaust velocity, would have a specific impulse of 850 seconds. A 'nuclear electric rocket', in which a nuclear reactor was used to produce electricity which propelled mercury ions to a higher exhaust velocity still, would have a specific impulse of at least 3000 according to calculations by the Los Alamos National Laboratory.

Several papers discuss the types of missions that might be flown to Mars. One of the most comprehensive is by Michael Tucker, Oliver Meredith and Bobby Brothers, of the Marshall Space Flight Center.

Firstly, they describe a mission using all-propulsive means of attaining martian orbit which might be flown in an opposition mission in 1999. At the rear of their spacecraft would be a rocket stage to propel the whole craft out of Earth orbit and on the trajectory to Mars. The motors for this stage would be derived from the SSME.

Next there would be another rocket which would place the craft into orbit around Mars and, at the end of the planetary excursion, propel it onto a trajectory to return to Earth. These motors would be derived from the OTV (orbital transfer vehicle) which is being developed for use with the Space Station.

The same engines would be used in a third stage which would brake the craft from its interplanetary velocity and place it in orbit round the Earth.

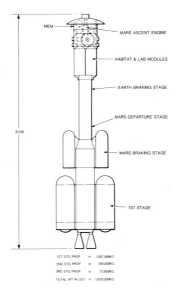

1ST STG PROP	=	1,027,000KG
2ND STG PROP	=	304,000KG
3RD STG PROP	=	72,600KG
TOTAL WT IN LEO	=	1,620,000KG

This Marshall Space Flight Center concept for a mission to Mars, aided by a gravitational boost from Venus on the outward journey, would use only retropropulsion, not aerobraking, to go into orbit round the Red Planet (NASA/MSFC).

At the forward end of the 'stack' would be the 'Mission Module', in which the crew would travel to Mars and conduct scientific activities. This would consist of three of the pressurized modules which are being built for the crew of the US Space Station. The MM would stay in orbit around Mars with a crew of two while the last element of the spacecraft, the Mars Excursion Module, descended to the surface with the other four members of the party. This would be the case with an opposition mission, but with the conjunction mission all six of the crew would descend to the surface.

The MEM would enter the martian atmosphere protected by a heat-shield with a diameter of 15.2 m, while three descent rocket engines would lower it through the last few metres. In this concept one of the three engines would be used to take the MEM back into orbit to rendezvous with the MM. Assembled in Earth orbit this spacecraft would be an impressive sight, 92 m long and with a mass of more than 1600 tonnes, of which the manned portion would be 132 tonnes. The manned spacecraft would weigh 173 tonnes on the three-year conjunction class missions, mainly because of the need to take extra crew systems and consumables.

The same manned craft could, however, be sent to Mars by a much smaller initial mass on an opposition mission in 2001 if aerobraking is used, as mentioned in the last chapter. In this case the first stage would be a rocket using LH and LOX in engines adapted from the SDV-3R and tankage derived from the Shuttle external tank. This stage would propel the spacecraft out of LEO but would then return to Earth orbit for further use.

On arrival in the vicinity of Mars the spacecraft, now consisting of a second stage plus the manned portion, would be braked into orbit

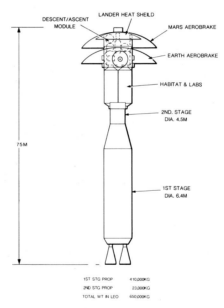

1ST STG PROP	410,000KG
2ND STG PROP	23,000KG
TOTAL WT IN LEO	650,000KG

This version of the Venus 'outward swingby opportunity' by Marshall Space Flight Center teams, would use aerobraking to go into orbit round Mars and also on its return to Earth. As a result of the saving of propellant its mass would be 650 tonnes instead of 1620 (NASA/MSFC).

using a heatshield with a diameter of 24 m. This heatshield is then either used to land the MEM on the martian surface or is jettisoned and a secondary heatshield with a diameter of 15 m used for this purpose.

After the landing the MEM returns to orbital rendezvous and the MM is propelled back towards the Earth by the second stage. Yet another aerobrake with a diameter of 24 m would be used to brake the MM into a highly elliptical orbit round the Earth, from which the crew would return to the Space Station with the aid of the OTV.

As noted by Sally Ride, the use of aerobraking reduces the LEO mass of a Mars expedition considerably—in this case from more than 1600 tonnes to 713 tonnes. Even greater savings would be realized by aerobraking for 1999 opposition (635 tonnes) and conjunction (581 tonnes) missions.

The authors illustrate two possible designs for a MEM. One would be based on the familiar Apollo command module shape, a cone with a rounded end and a diameter of 9 m, but this would be a dead end as far as growth was concerned. A design using a large heatshield would be preferable; it could be used for either large or small payloads.

These spacecraft concepts, plus one which used a hybrid plan, with aerobraking used at Mars but propulsive braking back in the vicinity of the Earth, would all involve the crew living in a state of weightlessness during the two interplanetary sectors and, for the crew who stay in martian orbit, another long period as well. The Marshall team explored another idea, artificial gravity, to ease the physiological and perhaps psychological problems associated with weightlessness.

The solution would be to rotate the whole spacecraft to create the impression of gravity. This would be done by placing the modules in which the crew would travel at the end of two openwork trusses each 60 m long. Rotating the whole craft four times a minute would then create at the ends of the trusses an artificial gravity of the same strength as on the surface of the Earth.

A similar solution, but with shorter booms of 37 m each, was proposed by Hubert P Davis, of Eagle Engineering Inc of Houston, who carried out 300 computer simulations of missions for NASA. Artificial gravity would be provided for two mission modules totalling 40 tonnes spinning around a central command module, also of 40 tonnes, which would be despun.

The design also included two Mars landers each of 75 tonnes, and four 'Mars Maneuvering Vehicles' to permit manned sorties from the parking orbit around Mars to Phobos and Deimos. The first stage of the vehicle, based on the SSME and Shuttle external tank, would propel the craft on a long, conjunction-style trajectory to Mars. A second stage would place the craft into martian orbit and a further rocket would send the command module back to Earth with the crew. Conjunction class missions can take a larger payload to Mars, and in this case the payload mass would be a remarkable 287 tonnes out of the total of 1250 tonnes.

Another design which shows ingenuity was prepared for the *Case for Mars II* conference of 1984 and is published in *Manned Mars Missions* by J R French of the Jet Propulsion Laboratory. His spacecraft would be a means of delivering crews and cargo to a Mars base on a regular two-year basis. It would consist of a three-armed structure, with two Space Station modules at the end of each arm.

The arms would be launched separately from LEO but would rendez-vous on the way to Mars and assemble themselves into a pin-wheel configuration which would rotate slowly to provide artificial gravity at the level to be found on Mars, i.e. one third of terrestrial gravity. Each of the twin modules would house five crew members who would transfer into 'Mars shuttles'—entry vehicles of a biconic shape which would descend to the surface. Meanwhile, the rest of the craft, called 'Deep Space Habitat' by French, would fire a rocket to send it back to Earth on a heliocentric orbit. There, it would be captured into Earth orbit for refurbishment and re-use. Each time it returned it would bring back to Earth crew members who had been on Mars for a two-year tour of duty. They would fly up to the Deep Space Habitat in the Mars Shuttles they had used to descend.

Barney B Roberts, of the Johnson Space Center, puts forward the interesting concept that the first manned mission to Mars might well be merely a 'fly-by' and not a landing.

This has history behind it, for the first two Apollo missions to the Moon, 8 and 10, did not involve landings. Although not strictly fly-bys, for the spacecraft went into lunar orbit on both occasions, they gave the programme invaluable experience without taking the final risk of landing.

The first Soviet approach to Mars may well be by means of a fly-by mission also. This, too, would provide experience before commitment to landing and the Roberts scenario may well interest NASA for reasons of expense and risk avoidance, especially as the mission would take only a year, reducing the problems of exposure of the crew to weightlessness and isolation. He postulates a spacecraft of only 324 tonnes, placed in LEO by the SDV class of launcher, as long as aerobraking can be used to achieve orbit on the return to Earth. The velocity of the spacecraft as it approached the Earth would reach more than $16 \, \text{km s}^{-1}$ and both the aerodynamic heating and the g forces might be excessive, the first for the vehicle and the second for the crew. If a rocket had to be burned to place the returning spacecraft into Earth orbit the resulting mass would reach 612 tonnes.

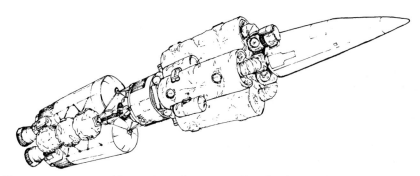

Five astronauts would travel in this 'Mars Shuttle' from low Earth orbit on a trans-Mars trajectory. A few days out, the shuttle would rendezvous with an interplanetary space station in a permanent repeating orbit between Earth and Mars and the biconic Mars Module would be docked to one of its three arms (Carter Emmart).

The spacecraft would consist of a mission module based on Space Station hardware, a three-man command module, and two rocket stages based on the orbit transfer vehicle.

The spacecraft would be aimed at a point about 300 km above the martian surface, where its velocity would reach $8 \, \text{km s}^{-1}$. The astronauts would be within ten planetary radii of Mars for a mere $2\frac{1}{2}$ hours, but there would be plenty of opportunity to make observations for days before and after the encounter. They would also be able to release probes to study the atmosphere and the surface. Doing away with the mission module and replacing it with a container for food and other supplies could bring the total weight in LEO down to no more than 210 tonnes, said Roberts.

Archie Young, Ollie Meredith and Bobby Brothers, of the Marshall Space Flight Center, point out in another paper on fly-by missions that a propulsive manoeuvre—a rocket burn—is necessary in such a flight to direct the craft back to Earth. Their mission would be launched on 9 March 2001, would reach Mars 172 days later on 20 August 2001, and

At a space station orbiting above the Earth, a manned mission to Mars is being assembled. The white vehicle at the top is a biconic Mars Module, the craft which will land on the planet. Beneath it is the 30-tonne Orbit Transfer Vehicle (OTV) which is used in the assembly process. In some schemes it would also form part of the Mars-bound spacecraft (NASA).

would return to Earth 442 days after departure, on 25 May 2002. Their 296 tonne craft would require only four flights of the SDV rocket and three Shuttles to place it into LEO. (The Russians would need only three Energiya launches for this type of mission.) It would have rocket engines of the OTV type and would be a mere 48 m long, which is less than twice the length of the Soviet Union's Mir space station together with its attached Soyuz TM spacecraft and Kvant scientific module, although it would have about ten times the mass.

In another paper Young shows that the use of Venus in a gravitational assist manoeuvre can be used at every opposition of Mars. Although this increases the payload which can be launched towards Mars it has the disadvantage that it increases the mission duration by 25 to 50 per cent. This may be more use for cargo vehicles in the distant future than for any manned flights.

Some of the papers give details of exotic types of propulsion which may one day be used to make the journey to Mars a much swifter business. A team from the Jet Propulsion Laboratory, Battelle Northwest laboratory at Richland, Washington, and NASA's Lewis Research

One of the two landing craft of a joint US–Soviet manned Mars mission prepares to descend to the surface, while the other waits in orbit. An imaginative concept of the future by Soviet artist Aleksandr Zakharov.

Center in Ohio, reported on a study of Pegasus (PowEr GenerAting System for Use in Space), an electric propulsion system using a multi-megawatt boiling liquid metal fast reactor.

The reactor employs a direct Rankine power cycle to generate 8.5 Mwe, of which 5 Mwe is used for the electric propulsion, either by means of a magnetoplasmadynamic thrust system or an ion thruster using mercury as the propellant. The electric propulsion systems give a far higher specific impulse than chemical propulsion and although the low thrust means a longer journey time this is more than compensated for by a reduction of the mass needed in LEO.

The authors found that a mass of 344 metric tonnes would be needed in LEO and the space vehicle would fly to Mars in 601 days.

There would be a 100 day exploration period and the return journey would take 268 days. Only 111 tonnes of propellant would be used, compared with the hundreds of tonnes in chemical rockets.

NASA is actively studying this aircraft concept for Mars exploration. The 20 m wingspan of the lightweight unmanned craft would give it good flying qualities in the thin martian atmosphere and it would carry a scientific payload of 45 kg to study the planet over a 4000 km range (NASA).

Even more exotic is the concept of propulsion by means of the annihilation of antiprotons. Particles of matter and antimatter mutually destroy one another when they meet, with the release of vast amounts of energy. Steven D Howe and Michael V Hynes, of the Los Alamos National Laboratory, point out that less than a gramme of antiprotons would be needed to provide the energy to propel a 400 tonne ship to Mars and back, using hydrogen as the propellant. The hydrogen would be heated to a high temperature and therefore a high exhaust velocity by the energy released by the annihilation of antiprotons in a rocket engine of the NERVA type. The specific impulse of such an engine would be 1100 seconds.

Howe also discusses the possibility of reviving the NERVA nuclear rocket programme. NERVA was a nuclear thermal rocket (NTR) which merely used the heat of fission to propel hydrogen propellant to a high exhaust velocity. With a specific impulse of about 900 seconds this rocket would like the other systems mentioned be able to reduce considerably, perhaps by two-thirds, the mass needed in LEO for a Mars mission. As suggested earlier, however, this suggestion is unlikely to be followed in the present anti-nuclear state of mind in the US.

Several specialists have suggested that the first Mars expedition may be to Phobos and Deimos, rather than to the planet itself. The Δv needed to land on one of these tiny moons once the spacecraft is in martian orbit is quite small, as long as the original orbit is chosen carefully. Phobos would be an admirable base from which to dispatch automatic probes to the martian surface and it also promises to be a valuable source of materials such as propellants, as mentioned in Chapter 4. Visitors would have to be careful how they moved, for the gravitational force on Phobos is so small that a really vigorous jump might launch a space-suited visitor into space. One moonlet mission which is being proposed is known as PhD, since it would involve visits to both Phobos and Deimos.

Whichever route is chosen for the first missions—landing, fly-by or moonlet—there does seem to be a historical inevitability about it. If only the US administration and NASA were involved this would probably not be the case, but the Soviet Union is rapidly building up the capability to perform one of the missions. At the first sign of a genuine commitment by the Russians there is bound to be a prompt response by America, either in competition or in cooperation.

One attempt to describe a manned mission to Mars which is particularly interesting because it is European is the work of Dr Harry Ruppe, one of the V2 scientists who worked closely with Dr Werner von Braun and then moved to the United States. He is now back in Germany as director of the Aerospace Department at the Technical University of Munich, but his scheme obviously relies heavily on American technology.

His plan requires a two-year preparation period in which the various component parts of the Mars vehicle are taken into low Earth orbit. There would have to be 35 launches of a newly developed Heavy Lift Launch Vehicle (HLLV) based on Space Shuttle propulsion technology, as well as five launches of the Shuttle itself.

At the end of the preparation period a 2000 tonne launch stage would be orbiting the Earth at an altitude of 325 km. It would consist of three large modules derived from the external fuel tank of the Shuttle, each about 31 metres long and 8.4 metres in diameter. The engines at their base would be improved versions of the main engine which powers the Shuttle. On 13 November 1996, these engines would ignite and place the whole 'stack' on a minimum energy trajectory to Mars.

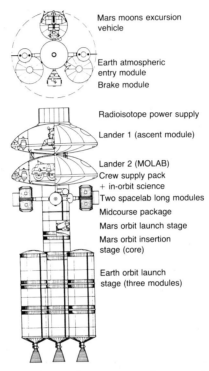

Mars moons excursion vehicle

Earth atmospheric entry module

Brake module

Radioisotope power supply

Lander 1 (ascent module)

Lander 2 (MOLAB)

Crew supply pack + in-orbit science

Two spacelab long modules

Midcourse package

Mars orbit launch stage

Mars orbit insertion stage (core)

Earth orbit launch stage (three modules)

Professor Harry Ruppe's design for a spacecraft on an initial Mars mission is both ingenious and complex. It would include two landers with aeroshells, one holding the ascent stage and the other a mobile laboratory (MOLAB) which would also be the living quarters at base camp (Professor H Ruppe).

Between the three modules there would be a huge rocket with a mass of 784 tonnes, whose purpose would be to decelerate the spacecraft so that it fell into orbit round Mars on 11 July 1997, after a voyage of 240 days.

Ruppe does not envisage the use of aerocapture or aerobraking but advocates the use of retropropulsion to place the whole spacecraft into a 1000 km circular orbit around the martian equator. This would mean burning 434 tonnes of propellant, all of which would have to be lifted out of Earth orbit.

Another rocket of 45 tonnes mass would be used to make a mid-course correction, necessary because of the inclination of Mars's orbit to the plane of the ecliptic. In addition to these rocket stages, the Mars manned mission spacecraft would have a mass of 1341 tonnes, of which the bulk would be made up of a 'crew supply package' of 790 tonnes, Mars vicinity equipment of 311 tonnes, and a 145-tonne rocket to launch the craft on its trajectory from Mars to the Earth.

The four-man crew would live in two lengthened modules based on the Spacelab hardware which has already flown on the Space Shuttle in Earth orbit. On arrival in Mars orbit a 63.5 tonne lander would be detached from the stack and descend through the atmosphere unmanned. Its speed would be reduced first by air resistance, just as the Apollo command module returned to Earth from the Moon. As the atmosphere becomes denser two parachutes would be deployed in succession to decelerate the lander further, and finally a soft landing would be achieved by last minute rocket braking. All these techniques or variants of them have been tried and tested in manned and unmanned landings on the Moon and Mars.

This first lander would contain a 12.8 tonne ascent vehicle, based on the Lunar Module which was employed in Project Apollo, and capable of taking the two-man exploration team back from the martian surface to the 'mother ship' in orbit. The crew themselves would not follow the first lander down until it had landed safely and signals indicated that the lander was intact and in working order.

Then they would follow in another 63.5 tonne lander, which would land in an identical way. Its payload would include a 45.2 tonne MOLAB (mobile laboratory) which would be both the headquarters of the ground crew and the method by which they would explore the planet's surface.

Since the Ruppe mission envisages a stay time of 550 days in orbit round Mars before conditions are favourable for return to Earth, the two men would have plenty of time for their exploration. Ruppe expects them to stay on the surface for 500 days, after a 25-day orbiting period to check the landers. After they had returned to the mother ship there would be another 25 days spent on checking the systems.

Ruppe has planned a testing schedule for the explorers. He even suggests that the MOLAB, powered by a radio-isotope device similar to those used in the Voyager project, could be used to drive from the

Equator to the South Pole, a distance of 5333 km.

This seems overambitious, for the average speed he envisages is only $7\,km\,h^{-1}$ and he proposes a drive to the South Pole lasting from the 50th day on the surface to the 245th. MOLAB would return to the base camp by the 440th day and the ascent stage would take off on the 500th.

At the end of their exploration, the crew would return to the orbiting mother ship, from which a crew member would in the meantime have journeyed in a special 12 tonne excursion vehicle to Phobos and Deimos. With the crew united again, the 230-day return to Earth would begin.

The Mars orbit launch stage would be ignited on 12 January 1999, and after a mid-course correction to change the plane of the trajectory they would arrive back at the fringes of the Earth's atmosphere on 30 August 1999. The only hardware which would then return to the Earth's surface would be a 6-tonne atmospheric entry craft based on the Apollo command module. The return from Mars would be faster than the similar return from the Moon, so a rocket would have to be fired to slow the module before it entered the atmosphere.

Dr Ruppe has developed a well thought-out mission scenario which skims over some of the big unanswered problems of martian manned missions, such as how humans will stand up both psychologically and physically to a total mission time of 1020 days.

However, it is a mission which lacks credibility, for it is far too ambitious for the early stages of manned Mars exploration. Its origins in Apollo technology are plain but Apollo is now more than 20 years in the past and much more appropriate technology is now available or soon will be. For instance, aerocapture can save anything up to 50 per cent in mass in low Earth orbit; this technology should be available by the late 1990s.

His estimate that the mission could be accomplished for a cost of $15 billion in 1984 dollars is also highly suspect. Apart from Project Apollo, which came in at very nearly the original estimate, the costs of space missions are usually underestimated by engineers, mainly because of unforeseen complications. Plans for a manned Mars mission will eventually be approved by some future Congress or Supreme Soviet, but it will much more austere than Dr Ruppe's.

7 The Human Factor

You can't disobey doctor's orders, can you?
Cosmonaut Aleksander Laveikin

On 30 July 1987, the Soviet spacecraft Soyuz TM2 landed in the steppes of Central Asia. As the tiny conical capsule approached the ground beneath its parachute, anxious doctors rushed forward with the ground recovery crew. As soon as the hatch was open they lifted out one of the three cosmonauts on board. It wasn't the first medical emergency the Soviet Union had experienced in space, but it was by far the most serious.

Aleksander Laveikin, a burly 36 year old Moscow-born technologist, had been a member of a two-man crew of the Mir space station since 7 February. Twice in the month of June he had left the Mir for EVAS (extra-vehicular activity, or space walks). On the first occasion, medical monitoring had shown that his heartbeat had become irregular.

Even so, he was allowed to embark on the second EVA a few days later. Doctors at the Mission Control Centre near Moscow were horrified when the cardiovascular irregularities appeared again.

Arrhythmias, or irregular heartbeats, had been experienced by Apollo astronauts during their Moon excursions. At the time the NASA doctors concluded that it was a lack of potassium in the diet that was affecting the heart muscle and some crews were forced to drink large quantities of an orange drink unappetizingly laced with potassium. But the problem may be more fundamental.

The heart muscle shrinks in weightlessness since it no longer has to counter the force of gravity in pumping blood to the upper part of the body, and all underused muscles shrink. It may be that the shrinking is associated with cellular changes which make the heart muscle less stable electrically and therefore more prone to arrhythmias. Whatever the reason, the Soviet doctors knew that they had to bring Laveikin back to Earth as soon as possible. Fortunately, the Soyuz TM3 craft was due to take a short-term crew of three to the Mir space station in July

1987. It was intended that Laveikin should stay for a full year in Mir with his partner Yuri Romanenko but his flight had to be cut short after 156 days.

This was a blow to the continuing programme by the Soviet Union to test the ability of men to exist in good health in space for protracted periods. At least Romanenko was able to continue with the flight, partnered by one of the crew of Soyuz TM3, Aleksander Aleksandrov, and set up a new record of 326 days. Apparently he was a fanatic adherent to the exercise programme laid down by the medical team for crew members on the Mir and it may well be that Laveikin was not quite so conscientious.

Whatever the truth of this possibility, and despite Western rumours to the contrary, Laveikin was able to return to full health within a short time of returning to Earth.

However, this episode is of great significance for any manned Mars expedition. Because spacecraft, like the Mir, in low Earth orbit are never more than a few hundred kilometres from Earth, it is possible to return a sick or injured astronaut to the surface within a few hours if necessary.

When an interplanetary spacecraft is launched from LEO at escape velocity, it is not so simple. There are certain 'abort modes', in NASA jargon, which would enable a Mars-bound ship to turn round and return to Earth. At first this would take days, later several months and ultimately it would not be possible at all, for to continue to fly by Mars and return to the neighbourhood of the Earth at the end of a year or more would be the quicker option.

In a typical mission described by Archie C Young, for instance, a quick return to Earth would be possible if the decision were taken in the first 1.75 days. Even then the return to LEO would take 18 days. The mission Young described depended on a swing-by manoeuvre at the planet Venus, and if it had to be aborted within the first 40 days the spacecraft could return to LEO in from 80 to 250 days. After 180 days the craft would have to continue to Mars and the return would take a total of 560 days.

Any medical condition or injury that required treatment of an urgent nature would perforce have to be dealt with on board the spacecraft.

The crew of any Mars craft will have to include at least one doctor skilled in surgery and several other people highly trained as paramedics to deal with any crises that were life-threatening or potentially so.

Even more significant to the prospects of a manned Mars mission than Laveikin's illness was a tragic sequel to the flight of Soyuz TM4 to the Mir space station in December 1987. The 'cosmonaut–researcher' on that eight day flight was Anatoli Levchenko, a healthy 46 year old test pilot of great experience who had not flown into space before.

His main assignment was apparently to conduct early tests of the Soviet space shuttle, and the Mir mission was mounted to give him some space experience before the shuttle flew. He duly returned to Earth on 29 December 1987, and appeared to be no more affected by the trip into space than any other cosmonaut.

Yet by 10 August 1988, he was dead, after what Moscow called 'a grave illness'. In fact, he died from a brain tumour but the flight was so short that any space effect can certainly be ruled out. But here was a case where a man who was cleared as fit to fly in December had died by the following August.

The seriousness with which the Soviet authorities viewed this event can be judged by their reaction in sending a doctor, Valery Poliakov, to Mir last September to monitor the health of Vladimir Titov and Musa Manarov, the two long-term crew members.

At some future date a crew member of a Mars-bound spacecraft might easily suffer the same fate as Levchenko after the craft had reached the point of no return.

Dr Philip C Johnson, of the Johnson Space Center in Houston, has pointed out that a previously undiagnosed malignant disease has time to become clinically significant during a trip to Mars. No-one thinks it unusual, he said in a paper presented to the Manned Mars Missions working group, when an individual develops clinical cancer a few months after being passed physically fit. This is exactly what happened to Levchenko.

Another matter pointed out by Dr Johnson is that over a period of years the radiation environment of space may lead to an increase in the rate of tumour formation because of its effect on the immune system. It might be necessary to choose middle-aged crews who no longer wished to produce children.

In the event of accident or disease, the prospect of surgery in a weightless state, incidentally, is a daunting one. To take only one of the difficulties, blood in an operation site will not ooze or flow conveniently into the surrounding cavity, from which it can be mopped up, as on Earth. It will migrate out of the wound as it is disturbed by the surgeon's activities and will form a red mist. Anaesthetic gases will probably have the effect of putting the surgeon to sleep almost as soon as the patient, so injected forms will have to be used.

105

And the lack of gravity will mean that an intravenous 'drip' to introduce fluids into the body will not work; it will be necessary to pump such fluids in.

A great deal will need to be known about the human response to weightlessness or, more accurately, microgravity, before it will be possible to launch towards Mars with full confidence in the ability of the crews to survive. The true state will be one in which there is a slight gravitational effect, caused by the mass of the spacecraft itself.

The problem of heart irregularities is only one of many which have cropped up in the last three decades as the scope of manned missions in space has expanded. The full year flown by Vladimir Titov and Musa Manarov in the Mir space station as a follow up to Romanenko's 326 days increased confidence but it is still nowhere near as long as the two years of an opposition mission or the three years of the typical conjunction flight to Mars. It does, of course, suggest that health problems on a sprint mission of the type advocated by Sally Ride would not be excessive as long as an unlucky selection like Levchenko was not in the crew.

There is no doubt that the steadily lengthening trips on the Salyut and Mir space stations have been deliberate steps on the road to learning the way to plan interplanetary trips. Enough Soviet officials have stated that this is so to make it certain.

The American case is more complicated, for the missions in their manned space flight programmes have all been short—the longest were the three Skylab excursions of 28, 59 and 84 days in 1973. Space Shuttle flights are never more than 10 days in length and it is clear that experience in extended flight will have to be earned in the Space Station, which will not be flying until the mid 1990s.

Sally Ride and other consultants have stressed that the Space Station programme must include intensive studies into life sciences. In her report she said: 'Without an understanding of the long-term effects of weightlessness on the human body, our goal of human exploration of the Solar System is severely constrained. Before astronauts are sent into space for long periods, research must be done to understand the physiological effects of the microgravity and radiation environments, to develop measures to counteract any adverse effects, and to develop medical techniques to perform routine and emergency health care aboard spacecraft . . . Until the space station is occupied, and actual long-duration testing is begun, we will lack the knowledge necessary to design and conduct piloted interplanetary flights or to inhabit lower-gravity surface bases.'

The first problem encountered by men and women in a weightless state is a feeling of disorientation and nausea, with actual vomiting in some cases. This is due to the fact that the vestibular apparatus in the ears, which is part of the balancing mechanism of the body, is no longer under the effect of gravity and the brain has no sense of 'up' or 'down'. The various signals received by the brain are contradictory and confusing. In short flights this 'Space Adaptation Syndrome' can be extremely serious for it can restrict the ability of the space travellers to perform necessary tasks.

It was not experienced in the first two generations of US manned space vehicles, Mercury and Gemini, which were both small. Apparently the spaceman needs to move about to a fair extent to experience it and the larger volume of the Apollo, Skylab and Space Shuttle gave them more opportunities to move their heads about sharply.

The acute symptoms do tend to disappear in only two or three days, but Soviet cosmonauts experienced a tendency in some cases for the disorientation to re-appear later in the mission. The syndrome is hardly likely to disrupt a two- or three-year mission to Mars, but space doctors will want to find ways of countering it if possible, if only to guard against sudden relapses at critical mission points.

Secondly, the general experience is that there is a loss of calcium from bones, since being in a weightless state is similar in some ways to prolonged bed rest. Calcium is laid down in bones from the blood-stream in response to electrochemical effects produced by the stresses placed on the bones in normal use. In a spacecraft in free fall the bones are on the whole not stressed and as a result the ordinary physiological losses of calcium from the bones are not made good. Soviet cosmonauts who have been on long space missions apparently recover their normal bone architecture after their return to Earth, but there is no certainty that the experience in one-year missions can be extrapolated to two- or three-year flights.

Exercise may be the answer, but quite rightly US experts question whether the prospect of four hours exercise a day would be appealing to astronauts on a long trip to Mars. Four hours vigorous walking each day is accepted as the optimum for avoiding calcium loss on Earth, which is why elderly people and others incapacitated and unable to walk far often show signs of this loss. From the point of view of a trip to Mars, there is more than one problem. After a year in free fall the deceleration needed first to place the spacecraft into orbit around Mars and then to land on the planet might be harmful to the weakened bones of the crew. There might even be fractures and then the astronauts would not be able to play their full part in surface activities. A protracted stay on Mars, necessary in the conjunction type missions, might be almost as harmful, since the force of gravity there is only one-third of the terrestrial force.

The crew would have to face more acceleration forces on lifting off from Mars and entering an orbit around the Earth—in the last case after anything up to three years in space. The need to restrict the g forces experienced on return to Earth may dictate the design of the spacecraft. It has been calculated for some missions that pure aerobraking would impose loads of up to $8g$ on the astronauts. This would be totally unacceptable and means that the velocity of the spacecraft would have to be slowed by retropropulsion before it hit the atmosphere.

The problem will have to be solved, especially as there is also loss of mass from muscles, including the heart muscle, as mentioned above. The period of recuperation after a three-year flight will certainly stretch into months, at least some of which may have to be spent in orbit round the Earth in a space station, where a gradual return to the harsh $1g$ environment of the terrestrial surface can be planned.

Another effect of weightlessness is a loss of red blood cells. The cause of this phenomenon is unknown, but it may be a facet of a general atrophy of tissue cells in response to weightlessness. The reduction in blood volume does not appear to be serious, but it is unknown whether the bone marrow, where the cells are made, would respond normally if an astronaut were to suffer haemorrhage from a wound or a bleeding ulcer.

Dr Johnson has identified several other health concerns. Infectious disease may strike a crew on a long space flight. This could only come, of course, from organisms taken onto the spacecraft by one of the crew or brought on in food or dust. Immunization or new anti-viral antibiotics would be a desirable item in the medical kit on a Mars-bound ship.

Apart from the possibility of osteoarthritis, mentioned above, middle-aged astronauts might find that the common condition known as presbyopia, 'old man's sight', will strike them suddenly. Previous flights have shown that weightlessness makes this condition worse, but it is easily corrected by reading glasses which should be taken on the trip.

Perhaps the most serious matter to be considered in medical terms is the radiation hazard to Mars-bound crews. All crews will be subjected to radiation in the form of electrons and protons as they pass through the Van Allen radiation belts that circle the Earth, although this area is passed through so quickly that it is not thought to be a major hazard. More prolonged will be the exposure to the galactic cosmic rays that pervade interplanetary space. Finally, there will be the most serious risk of all from solar flares or 'Solar Particle Events', which, unlike the other radiation levels, cannot be predicted far ahead. An SPE of unusual proportions occurred in August 1972.

If a similar flare occurred during a manned mission to Mars the crew would receive a dose which would increase their risk of developing cancer during their lifetimes by a significant percentage. This risk, which is twice as high for women as for men, could be considerably reduced by supplying the crew with a 'storm shelter' into which they could crawl for the duration of the storm of particles expelled by the Sun. The shelter would be more effective the thicker its walls— aluminium of $15 \, \text{g cm}^{-2}$ would reduce the risk by an order of magnitude over $2 \, \text{g cm}^{-2}$, which is more typical of space vehicles.

Another possibility is the provision of a shelter with walls containing water, which would also shield against the high-energy particles of an SPE.

The tenuous atmosphere of Mars would also have a beneficial effect if the flare should occur while the crew were on the martian surface. In addition, for half the time each martian day the bulk of the planet would act as a shield. Any habitat which is established on Mars should ideally be covered with a minimum of 10 cm of martian soil to provide more protection against this and other radiation hazards.

The conclusion of Dr Stewart Nachtwey, of the Johnson Space Center, is that radiation hazards will not forbid manned Mars missions, but will have to be considered in the design and operation of the spacecraft and of the base on the planet.

The radiation dose received by any crew member of a Mars mission would be carefully monitored and if it were higher than some agreed minimum for a particular member that member would probably not be allowed to fly into space again. This would be similar to the situation in which an industrial radiation worker who exceeds the life-time dose laid down by international or national standards is removed from work in which he may be exposed to radiation, and the dose laid down for travellers to Mars will probably be similar.

Another health problem is generated by the fact that, intrinsic in the design of a spacecraft bound for Mars, will be the inclusion of many man-made materials which 'out-gas' with age. This aspect will have to be very carefully monitored, because the Mars craft will be a closed system—any toxic gases which are emitted by materials will accumulate in pockets and will not be stirred up by convection currents, which do not occur in the absence of gravity.

Dr Martin E Coleman, of the Johnson Space Center, has listed a frightening catalogue of substances that have outgassed from spaceflight hardware materials in tests. They included acetaldehyde, acetone, methylethyl ketone, isopropyl alcohol, xylene, trichloroethane, toluene, methane and Freon 113. Metabolic products excreted by crew members included carbon dioxide, carbon monoxide, pyruvic acid, methane and skatole.

There was one near-tragedy when the Apollo spacecraft which linked up with a Soviet Soyuz craft in 1975 returned to Earth. A toxic gas entered the capsule during the landing phase and the astronauts, Tom Stafford, Deke Slayton and Vance Brand, were seriously affected. The incident might have been even more serious if they had had to breathe in the fumes for more than a few minutes before the hatch of the capsule was opened.

Mars-bound crews will not have this option, although they may have the option when a serious build-up of toxic gases occurs of donning their spacesuits for a period and venting the spacecraft's atmosphere. The craft will have to be equipped with alarm systems, protective clothing and real time atmospheric analysers to detect contamination.

Perhaps as important as the physical effects of flying to a distant planet are the possible psychological effects. There are some analogues of a manned Mars mission in the history of exploration. Sailors who embarked with Captain Cook, for instance, knew that they would be isolated in their craft, smaller than the ships which will take men to Mars, for two or three years as he explored the Pacific. There was always the chance that they would encounter other people, which will be denied to the Mars explorers, but the isolation from home and civilization must have been even more profound. With no means of communicating with home they were entirely cut off from their world.

The scientists who spend months on end in the Antarctic bases are less cut off, since at least they can speak to their families by radio. On the other hand, nuclear submarine crews who are completely cut off from the surface for 90 day missions must experience some of the feeling of isolation that will be experienced by Mars explorers. But the space travellers will be the first to remove themselves completely from the influence of the Earth—the Apollo astronauts were never more than three days away from splashdown if there were an emergency.

This isolation, especially if there is a mission in which the spacecraft is on the opposite side of the Sun from the Earth and radio communication is interrupted, will be the most profound ever suffered by humans.

Patricia A Santy, of the Johnson Space Center, lists these symptoms which may be the result of isolation: boredom, restlessness, anxiety, sleep disturbances, somatic complaints, temporal and spatial disorientation, anger and an inability to perform tasks as well as previously. These have been consistent findings in studies of submariners, explorers and volunteers in isolation experiments. Santy points out that compatibility between individuals with diverse backgrounds will be essential in selecting a crew. Sexual rivalry may be a problem in a mixed crew, perhaps leading to jealousy and tension; the answer may be to send married couples.

The 'right stuff' for a Mars mission will probably be people who would have been entirely the wrong stuff for the first US space flights. The first astronauts were mature men but they were test pilots who were of a temperament needing constant stimulation. They would be among the first to show the symptoms of boredom and frustration cooped up in a tiny cabin for months on end. Mature, contemplative people with a creative interest in something like painting or music, preferably over 45 so that they had no young children, may make up the first Mars crew.

Some authorities have insisted that the crew should be odd in number—an even number may split up into two equal factions which could be the cause of quarrelling. One possible unfortunate result of the 'pin-wheel' configuration of the *Case for Mars* spacecraft mentioned earlier is that the three sections of the crew will be isolated from each other. It is the nature of humans to identify with a group and the three sections of the crew may see themselves as rivals after months of isolation.

Even merely having a strong commander may not solve all the problems. In the history of the human race it is not only Captain Bligh of the *Bounty* who has deteriorated into a despot over a tightly enclosed group of subordinates. An unusually well balanced individual with great insight into his own and others' psychological make-up will be needed as commander if mutiny is not to threaten.

Alternatively, it may be desirable to replace command with a democratic form of management in which everybody has a voice. Both the American and Soviet programmes have occasionally exhibited a form of 'them and us' antagonism, in which the men and women in space have rebelled against what they consider to be unreasonable requests for activities or information. This arises from a feeling of isolation and the sensing that the people on the ground are living comfortable, routine lives, while 'up here' the astronauts are on the frontier. This experience may be even more marked as the Mars crew get further and further from Earth and direct conversation is made impossible by the lapse of time between the transmission of a signal and its reception at the other end of the link. Great tact will be needed by the ground controllers.

Despite all these caveats, the medical and psychological experts who have studied the problems associated with manned flight to Mars have concluded that there is nothing in the human mind or body that will forbid it. Men and women can and will take part in such missions and although there may be casualties on the way they will not be at a level which is considered unacceptable, just as in the past many terrestrial explorers have died but exploration was not inhibited.

8 The Cost

A Mars mission would require great expenditure which would only be justified in the event of automatic devices exhausting their capabilities.
Academician Roald Sagdeyev

The severest hurdle which the planners of a manned mission to Mars will have to overcome, over and above the technological advances that have to be made, will be budgetary. The problems of poverty, inner city decay, crime, Third World hunger and environmental deterioration can to some extent be alleviated by injections of cash, and there will undoubtedly be fierce resistance to the expenditure which will be necessary to send men to Mars.

This will be true in both the United States, which has a severe budgetary deficit problem which is not likely to disappear quickly, and the Soviet Union, where the much heralded *perestroika*, or reconstruction, will make demands which will not all be met. The third possibility, of a collaborative effort, say between the US and the Soviet Union or even the US and Europe, is so hypothetical that I think it can be disregarded, at least for the foreseeable future. There are too many technical, political and fiscal problems for such a venture to be readily accepted, although Harrison H Schmitt, the Apollo 17 astronaut who later became a US senator, has put forward the idea that an organization called INTERMARS should be formed to plan and carry out Mars missions.

It would have the same form as INTELSAT, the multi-nation organization that runs the worlds communications satellite system. All nations and organizations which wanted to belong would share in the costs and the benefits. It may come, but probably not in this century.

So how much would a preliminary mission cost, and how would it be financed? It is possible to state straightaway that no private venture capital will be involved—all the money will have to come from governments as a deliberate political decision. There already are private enterprise projects in space, particularly in telecommunications, and there will be more, when industrial processes are conducted in the NASA Space Station. In the distant future there will certainly be private mining projects to bring back the valuable minerals which are to be found in the asteroids, but there will be no profits emanating from Mars for decades, and perhaps centuries.

In assessing how much the exploration of Mars will cost it must be accepted that certain facilities exist which will be essential in mounting the first expeditions, and others will exist by the time they begin. Therefore, it might be argued that part of the cost of those facilities must be included in calculating the expense of going to Mars. Alternatively, if you are a keen advocate of manned exploration you can discount these costs, claiming that they would have been incurred in any case.

The most important of these facilities is the Space Station, which appears to be essential for the types of missions which have been proposed by US enthusiasts.

Exactly how the Space Station would be used in a manned Mars project remains to be decided. There are some elements in NASA who believe that it would be impossible to combine the support of a Mars project with scientific and industrial activities in a Space Station. The assembly, check-out, fuelling and launch of a multi-module Mars ship in and around a Space Station would, they say, so distort the immediate environment that in essence the station would have to be entirely devoted to fitting out the expedition.

Propellants, combustion products and other gases would, for instance, surround the station for months or years on end, so degrading observations by astronomical instruments. The spatial attitude of the station might have to be dictated by the requirements of the Mars travellers and its orientation to conduct Earth observations would thus be compromized. Assembly of modules in or very close to the station would create vibrations which would effectively render many scientific measurements impossible.

If these criticisms turn out to be true the whole cost of the Space Station might have to be borne by the Mars project. The same would be true of a Soviet venture if their version of the Space Station had to suspend all other activities while the expedition was prepared.

The Heavy Lift Launch Vehicle (HLLV) is another element in the plan which could either be costed as part of a US Mars expedition or excluded on the ground that it was needed for other aspects of the space programme as well. In similar fashion, should the development costs of the Soviet Energiya rocket, already incurred for Earth-orbiting programmes, be debited to the Mars plans, at least in part? The accounting methods used in the Soviet space programme are unknown but it is certain that there must be some form of control of the purse strings, otherwise the demands on the Soviet economy by the enthusiastic space community there would be unending.

It is in this context of flexible budgetary decisions that it possible to be not too critical of an estimate made by Professor Harry Ruppe, mentioned in Chapter 6, without accepting it as reasonable.

He assumes that the Space Station and the HLLV are already developed and operational. He then estimates that the total cost of his baseline mission would be $15 billion in 1984 dollars. This is broken down as follows:

Development, including Mars lander and MOLAB	$5b
Procurement of hardware	$2.5b
Transport to LEO	$3b
Orbital assembly	$1b
Flight control	$1b
Science programme	$1.5b
Contingency	$1b.

Needless to say, this level of expenditure is not generally accepted in the space community. It is certain that even with accounting procedures allowing the cost of existing facilities to be written off against other programmes the manned exploration of Mars will cost more than $15 billion per mission.

In her report to NASA, Sally Ride did not go into the dollar cost of the sprint missions she proposed. However, she does touch on the effect of her proposals on the level of NASA budgets. Specifically she says:

> Exploring, prospecting and settling Mars are clearly the ultimate goals of the next several decades of human exploration. But what strategy should be followed to attain those goals?
>
> Any expedition to Mars is a huge undertaking, which requires a commitment of resources which must be sustained over decades. This task group has examined only one possible scenario for a Mars initiative—a scenario designed to land humans on Mars by 2005. This timescale requires an early and significant investment in technology; it also demands a heavy lift vehicle, a larger Shuttle fleet, and a transportation depot at the Space Station near the turn of the century.
>
> This would require an immediate commitment of resources and an approximate tripling of NASA's budget during the mid 1990s.

In this context it is worth recording that the NASA budget for the fiscal year 1989 was $11.5 billion, but even tripling this would make it no more than ten per cent of the US defence expenditure. Ride continues:

> More important, NASA would be hard pressed to carry the weight of this ambitious initiative in the 1990s without severely taxing existing programmes. NASA's available resources were strained to the limit flying nine Shuttle flights in one year. It will be difficult to achieve the operations capacity to launch and control 12 to 14 Shuttle flights per year and assemble, test and continuously operate a Space Station in the mid 1990s. It would not be wise to embark on an ambitious programme whose requirements could overwhelm those of the Shuttle and Space Station during the critical next decade.
>
> This suggests that we should revise the ground rules and consider other approaches to human exploration of Mars. One alternative is to retain the scenario developed here, but to proceed at a more deliberate (but still aggressive) pace, and allow the first human landing to occur in 2010. This spreads the investment over a longer period, and though it also delays the significant milestones and extends the length of commitment, it greatly reduces the urgency for Space Station evolution and growth, and consequently for transportation capabilities as well.

A more detailed study of the cost of landing men on Mars was undertaken by Humboldt C Mandell, of the Johnson Space Center, although its date, 1981, makes it a little difficult to relate it to present day circumstances.

He said that although much of NASA's current planning was based on the perception that manned planetary flights would be more costly than the Apollo project, landing an American on Mars would actually cost only one third to two thirds as much as the lunar landing. He asserted that Apollo cost the American public $325 per head, whereas a Mars project would cost less than $200, both figures in 1981 dollars. Inflation since 1981 and in the years to come would presumably not alter this relationship in his estimation.

He mentioned the EMPIRE (Early Manned Planetary and Interplanetary Excursions) studies undertaken by NASA in the early 1960s. At that time Apollo was costing annually about $9 billion at 1981 prices, but budgets began to decline and severely limited the scope of future programmes. Senior policy makers had their minds set on the idea that manned planetary flights would be far more costly than Apollo. Mandell claimed that in fact two major changes since those early days would make it possible to reduce the cost of manned exploration.

The first of his claims—that the proposed sizes and masses of many of the key sub-systems in a Mars spacecraft have been reduced—still holds good.

However, the second claim—that the advent of the Space Shuttle has substantially reduced the cost of transportation to low Earth orbit—has been proved false. The slow turnround of shuttles, the inability of NASA to launch as many times per year as planned, and above all the disaster to the Challenger in 1986, have all meant that the USA has failed to bring down this cost.

Mandell made cetain assumptions in calculating the cost of a mission. He took as his baseline a 1971 NASA study, *Manned Mars Exploration Requirements and Considerations*, which assumed a 600 day mission, with ten astronauts in two completely independent spacecraft, each with a mass of 1770 tonnes. The mission was also assumed to use solar power, artificial gravity, assembly in low Earth orbit, and aerocapture at Mars. The mission would need about 100 launches of the Space Shuttle.

On this basis, and noting that in 1981 dollars Project Apollo cost $63 billion, Mandell estimated that the Mars mission would cost just under $20 billion without weight growth, and $23 billion if the weight grew by 33 per cent. The 'high probability range' of the cost would be between $20 and $40 billion. Over the 12 year length of the project the cost in 1981 dollars would never rise above $3.5 billion, whereas on the same basis the actual cost of Apollo peaked above $9 billion and for five years in the late 1960s never fell below $7 billion per year.

Finally, the total cost of the Apollo programme represented 2.8 per cent of a single year's GNP. Mandell estimated that the Mars project would cost only 1 per cent of the GNP of 1990, which will be three times as great as the 1969 GNP in real terms.

Mandell states: 'With relatively low risk, and building on existing space technologies, ten Americans could be sent to the planet Mars within twelve years from go-ahead. In our lifetimes there is only one such achievement remaining for mankind. It seems obvious that programme cost should not be a deterrent. The implications for the future of the nation, economic, social, political and motivational, are profound.'

As an employee at the Johnson Space Center Mandell was of course not entirely unbiased in his calculations, this being where much of the work of designing and developing a Mars ship would be carried out. The years that have passed have tended to remove some of the glamour of spaceflight and even the Space Station is facing budgetary hurdles that will probably delay its full implementation.

Nevertheless, the general outline of Mandell's paper can be accepted as a reasonable analysis of the relationship between Apollo and Mars manned landing costs. Time will tell whether administrations in the 1990s will be prepared to triple the NASA budget, as Ride suggested.

The best hope for those sections of American society who wish to see manned exploration of Mars is that the East–West disarmament agreements progress to the point where defence budgets are genuinely reduced.

In the summary report of the 'Manned Mars Missions' working group, the editors produce another series of estimates. Six different cases were studied and they were all found to be in the range of $24 to $28 billion in 1985 dollars. It was also found that the cost of a single manned mission to Mars would not be significantly different from the cost of Project Apollo up to the first lunar landing. Even with the construction of the space station and an SDV rocket, it is concluded that this mission could be achieved with only a three per cent annual increase in real terms in the NASA budget.

The major inducement to a Congress reluctant for two decades now to splash out large sums on manned spaceflight would be an overt statement by the Soviet Union that their own manned mission was being prepared. This would be just the spur that was needed, although it is doubtful if a renewed 'space race' would provide the right atmosphere for such a serious undertaking.

In the 'Manned Mars Missions' working group report the cost study was by Joseph Hamaker and Keith Smith, from Marshall Space Flight Center, who examined the potential costs of several options.

Since they used 1985 dollars as their basis it is not easy to relate their estimates to Mandell's, but they are certainly roughly in the same range. In drawing up their estimates, they made certain basic assumptions which should be listed.

One was the existence and availability of the SDV launch vehicle mentioned in Chapter 6, in particular, the form known as the SDV-3R. This would be a rocket based on Shuttle components with three Space Shuttle main engines in a recoverable pod (hence the 3R). The SDV-3R would be a highly capable rocket with a payload capacity to low Earth orbit of 82 tonnes and the authors of the report say that its existence is perhaps the most debatable premise in their estimates.

Other requirements would be the presence of the Space Station in LEO, an orbital transfer vehicle, an orbital manoeuvring vehicle, and advances in the design of rocket engines and deep space communications. The OTV, which is being developed by NASA, is a free-flying 30-tonne space vehicle to be used for transporting payloads between different Earth orbits. The OMV is a smaller craft to be used for moving payloads and spacecraft components about in the vicinity of the Space Station. The following cases were studied.

1. A 1999 opposition mission, using only rocket power to brake into martian orbit, fuelled by liquid hydrogen and oxygen. Excluding the cost of the SDV-3R rocket, which the authors believe would be used for other missions as well, this would cost an estimated $26 843 000 000.
2. A 1999 conjunction mission, also using only LH/LOX rockets; cost $25 877 000 000.
3. A 1999 opposition mission, using aerobraking to go into martian orbit; cost $23 792 000 000.
4. A 1999 conjunction mission, using aerobraking; cost $23 902 000 000.
5. A 2001 opposition mission, propulsively braked; cost $23 365 000 000.
6. A 2001 conjunction mission, aerobraked; cost $23 162 000 000.

The slight differences in cost, never more than a few per cent, are explained by the lower weights which are needed for aerobraked missions and for conjunction missions, in which the propulsive requirements are less. Hamaker and Smith conlude that, because of the cost and technical requirements, the most attractive option is the sixth, the 2001 conjunction mission with aerobraking. After the passage of four years, with still no commitment from the US Government to a manned mission, this particular flight is out of the question. The difference between the aerobraking and propulsion missions remains good, however, for future launch windows.

The main cost of the 2001 mission would have been $11.201 billion for the spacecraft in which the crew would travel to Mars and land on the surface. The rocket stages to propel the craft from LEO to and from Mars and to land on the surface would cost a further $5.629 billion and the balance would be made up of launches from Earth on the SDV-3R, scientific experiments, launch facilities, mission control and training. The authors estimate that launch vehicle development would cost a further $4 billion.

They do some creative accounting in their attempt to relate the cost of the 2001 mission to the historical cost of Project Apollo and conclude that in broad terms the two programmes would be comparable. They do this by taking the cost of Apollo up to the first landing on the Moon and eliminating some costs such as those of the Saturn 1 rockets which were precursors to Saturn V and never flew to the Moon. In 1985 dollars the Apollo cost would then be about the same as the 2001 conjunction mission to Mars.

There are clearly two sets of people in the world as far as Mars missions are concerned. They are not equal in size, for the vast majority of the people of the world would not be able to justify the cost of such missions. However, the small minority of dedicated scientists and engineers who believe in the inevitability of manned exploration of the planet feel that such expenditure is justified.

Their case is that the human race will eventually go to Mars at some time in the future, and it is better to go there as soon as the capability exists rather than to postpone the decision and go at some indeterminate time in coming centuries. It is impossible to be neutral in this debate and my own position is that if we wait until all the world's problems are solved we will never go. There will always be problems on the Earth's surface, but if the argument that expensive research should not be conducted until they are all solved had been followed in the 1950s the extremely valuable benefits of space would not have accrued in the 1980s.

As a footnote to the cost argument, it may be mentioned that the respected Stockholm International Peace Research Institute (SIPRI) has estimated that the Iran–Iraq war cost the antagonists more than $200 billion. The two nations could have financed several Mars missions and still have been better off! Total global military spending is now running at more than $1000 billion per annum.

Technology already exists, or is close at hand, that is capable of supporting manned flight to Mars and few would argue that military spending on this scale is more desirable. But only when the first missions have been successfully completed will it be possible to embark on the next phase of man's expansion through the Solar System—the colonization of Mars.

9 The Transformation

The turn of the Third Millennium presents an increasingly responsive environment for young men and women from all nations to join in an enterprise unique to our times: a project to establish a permanent human outpost on Mars by the end of the first decade of the new millennium.
Harrison H Schmitt, Apollo 17 astronaut

It is an inescapable fact that Mars is a cold, dry desert with an atmosphere that is completely useless to an air-breathing species. It is also many millions of miles from Earth and there is no prospect that large numbers of people will be going there in the near future. So what valid reasons other than blind faith can be put forward to justify the claim that Mars will be the next home of humanity? Why discuss a feat of planetary engineering that seems as distant to us as a trip to the Moon must have seemed to people living in the 19th century? In the words of the oft quoted justification for climbing Mount Everest: 'Because it's there'.

There is plenty of evidence that NASA is taking very seriously the idea of landing on and colonizing Mars, for example, this speech made by Mr James Fletcher, NASA's Administrator.

> The present martian climate would be modified through the melting of its northern polar ice cap. The cap would be melted by scattering carbon black pigment over its surface to enable the solar rays to be absorbed, rather than reflected.
> This melting would create a runaway greenhouse effect in which carbon dioxide and water vapour would be released into the atmosphere to raise the planet's surface temperature.
> Once the planet has warmed, certain organisms, such as the blue–green algae found on Earth, could be planted on the surface of Mars.

Mr Fletcher then questioned how much energy would be needed for this process, and how long it would take. He went on:

> While there are many unknowns and uncertainties which must be clarified before these questions can be answered accurately, rough

calculations by NASA scientists indicate that to carry out step one would require about 3×10^{23} calories. This is about the amount of solar energy falling on Mars in about 6 years. If we were to capture 1 per cent of Mars's solar energy to warm the planet it would take 200 to 300 years.

The second step—creating a breathable atmosphere similar to that at sea level on Earth—would require about 40 times as much energy as the first and could take 40 times as long, or from 8000 to 12 000 years. But that estimate could be cut in half simply by using the abilities of humans on Earth to survive in rarefied atmospheres at very high altitudes.

Of course, these processes also could be speeded up as a result of advances in science and technology that we can hardly imagine today, advances that would give us more knowledge about human physiology, planetary ecology and planetary atmospheres. NASA is working in all these areas today.

Attempting to make Mars habitable would be a project of enormous scope that could indeed take many centuries. But one of the major goals of a manned expedition to Mars could be to start a pilot project of this kind on a very small scale, to be followed by more extensive efforts by permanent residents in the future.

The challenge would be great, but the rewards could be even greater: nothing less than a new home for humankind.

Although, as discussed earlier, Mars is so much smaller than the Earth, its land area, in the absence of oceans, is almost exactly the same as the land surface of our home planet.

The record of the past shows that new technology inspires new outward movements of peoples. When, towards the end of the 15th century, reliable ocean-going sailing ships were being built, a movement began which in three centuries saw Europeans discovering and settling new lands in the whole of the Americas and Australasia, as well as dominating much of Asia and Africa.

The descendants of the European colonists still possess most of their gains of the centuries from 1500 to 1800. We still do not know the precise technology which will conquer Mars, but the prize will be enormous—millions of square kilometres of virgin land, all of it within the boundary of the 'ecosphere'—the hollow sphere surrounding the Sun and containing the orbits of the Earth and Mars which is the only part of the Solar System where liquid water and therefore life can survive.

Apart from the questions of temperature and atmospheric content and pressure Mars does not seem a natural home for life. For instance, soil is an important consideration as well. Soil is not merely inorganic matter weathered down from rocks; it is, at least when fertile, a teeming mass of organisms—bacteria, algae, fungi, protozoa—all of which are essential to the growth of plants. Most plants, for example, cannot fix their own nitrogen from the air and employ bacteria to do it for them. Many trees rely on underground fungi associated with their roots to break down organic material in the humus so that the roots can absorb nutrients.

So however favourable other factors might be, terrestrial plants transposed to the martian surface would grow only with difficulty. How is this problem to be overcome?

The most comprehensive NASA study of this problem was carried out by a team of scientists coordinated by the Ames Research Center in California. Entitled *On the Habitability of Mars, An Approach to Planetary Ecosynthesis*, it embraces all aspects of 'terraforming' and 'planetary engineering'.

Its final conlusion is that 'no fundamental, insuperable limitation of the ability of Mars to support a terrestrial ecology is identified'.

It goes on: 'The lack of an oxygen-containing atmosphere would prevent the unaided habitation of Mars by man. The present strong ultraviolet surface irradiation is an additional major barrier. The creation of an adequate oxygen and ozone-containing atmosphere on Mars may be feasible through the use of photosynthetic organisms.'

Some basic facts about the martian environment are adduced by the study. For instance, it is estimated that there are roughly 2000 hours in each martian year in which the surface temperature exceeds 270 K, which the report describes as 'a typical lower limit for growth of most terrestrial organisms'.

Comparing aspects of the environment with terrestrial equivalents, the study points out that Mars is an environment hostile to biology. Only anaerobes, generally single-celled organisms, would grow in the lack of oxygen; ultraviolet light, particularly in the range of 2000–3000 Å, would reduce the mean survival times of organisms to a few minutes at most, and the temperature range would allow seasonal growth for only 7 hours a day at the equator and mid-latitudes. The report goes on:

> Other important chemical elements for life, such as nitrogen and phosphorous, have yet to be identified on Mars. Even a most optimistic appraisal suggests that the kinds of terrestrial organisms able to survive in the present martian environment are quite limited, and the growth of

123

even these forms would be quite restricted in vigour and extent.

There have been many attempts to determine the response of microorganisms to simulated martian enviroments. While conclusions have varied, in part a reflection of the experimenters' choice of organisms and environmental conditions, investigators feel that there is a definite possibility for growth of certain anaerobic, cold-adapted terrestrial bacteria on Mars.

A comparison of the four photosynthetic groups of organisms that exist in the dry Antarctic valleys, the nearest Earth environment to martian conditions, shows that lichens and blue–green algae come closest to an ideal Mars organism. Lichens have an extremely slow growth rate and would therefore be unsuitable as a provider of an oxygen atmosphere. The blue–green algae are sensitive to UV, but as they can occupy subsurface niches they may be protected from this martian problem.

There would also be a layer of dessicated cells on the surface of an algal mat, which would help to reduce water loss.

Computer models reported in the Ames study indicated that blue–green algae covering a quarter of the martian surface for 7000 years would generate 5 mb of oxygen, about the same amount as the present carbon dioxide atmosphere. The minimum necessary for human breathing, about 100 mb, would take 140 000 years.

These are not compatible with the figures given by James Fletcher, but his results assumed that the greenhouse effect would be used before the algae were planted.

The Ames report discusses the greenhouse effect and points out that water vapour is much more effective than carbon dioxide in bringing it about. If enough water vapour could be added to the atmosphere to increase the total pressure by only 10 per cent, the resulting average temperature rise would be 10 K, compared with 7 K if the northern polar cap were all carbon dioxide and was all vaporized.

There could be a stable climate regime with a polar surface winter temperature of 190 K and a surface air pressure of approximately 1 bar. This could be brought about by spreading sand or dust on the polar cap and so reducing its albedo for about 100 years. The release of carbon dioxide would lead to advective heating, in which heat is transmitted from low to high latitudes.

The resulting polar heating would release more water vapour which would provide a stronger greenhouse effect. The higher temperature would enormously increase the area of the planet available for growth and this, combined with genetic engineering to produce species which would find martian conditions more congenial, would transform the prospects for terraforming Mars.

The Ames report dates from 1977 and it is no exaggeration to say that there have been enormous leaps in genetic engineering since then. The position, nature and function of many more genes have been identified and the tools which are used in the science have been refined. Another twenty or thirty years must bring a real possibility of genetically engineering 'martian' organisms which would thrive there. One possibility mentioned by the Ames report is that a species could be manipulated in such a way that it became more resistant to UV by acquiring an ability to repair damaged cells.

The report adds: 'In principle, the entire gene pool of the Earth might be available for the construction of an ideally adapted oxygen-producing photosynthetic martian organism.'

On the Habitability of Mars discusses all aspects of the terraforming and planetary engineering of Mars at length, particularly the use of blue–green algae and lichens to form an oxygen atmosphere, and comes to a number of conclusions.

They can be summarized as follows.

The lack of an oxygen atmosphere prevents Mars from being inhabited by man. The diurnal temperature fluctuations, intense UV radiation and dust storms can be dealt with by adequate shielding.

Lack of oxygen will not prevent the growth of all terrestrial organisms.

Blue–green algae and lichens could survive and grow on Mars and produce oxygen slowly, but this rate could be increased dramatically by genetically engineering species to adapt them to the martian environment.

Altering either the martian environment or available photosynthetic organisms, or both, would significantly decrease the time required to create an acceptable human habitat on Mars. Indeed, it may be mandatory to do so. If these steps are taken, Mars may well be made into a habitable planet.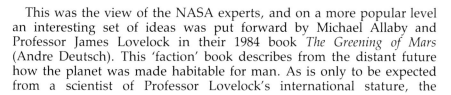

This was the view of the NASA experts, and on a more popular level an interesting set of ideas was put forward by Michael Allaby and Professor James Lovelock in their 1984 book *The Greening of Mars* (Andre Deutsch). This 'faction' book describes from the distant future how the planet was made habitable for man. As is only to be expected from a scientist of Professor Lovelock's international stature, the science is impeccable, but the technology which is described in the book is seriously flawed.

The most serious flaw is in the method used to raise the temperature of Mars. As described in the book, something like 5000 redundant solid fuelled intercontinental ballistic missiles are used to carry the warming agent to Mars.

Chlorofluorocarbons, or CFCs, are gases which are used to propel aerosol can contents and for heat transfer in refrigeration systems. They are widely believed to be responsible for damaging the Earth's ozone layer, since the chlorine they contain is said to deplete the ozone by combining with oxygen atoms.

The Allaby–Lovelock idea is that after strategic arms limitations agreements have been signed the American and other Western solid-fuelled missiles such as Poseidon, Polaris and Trident would be used to take CFCs to Mars. CFCs are a thousand times more efficient than carbon dioxide in bringing about the 'greenhouse' effect and Allaby and Lovelock postulate that a beneficial rise in the martian temperature could be brought about in a fairly short time by crashing rockets with CFC payloads onto the surface.

Trident rockets cannot reach escape velocity, so four rockets would have to be fastened in a cluster to form a first stage, with a fifth rocket bearing the payload as the second stage. This, of course, immediately reduces the number of payloads that are to crash into Mars from 5000 to 1000, each with perhaps 1 tonne of CFC.

It is impossible to say whether 1000 tonnes of CFC would have the desired effect but it seems doubtful to say the least. Even so, there are a number of more fundamental objections.

Firstly, not every launch window is equally favourable and when Mars is furthest away at opposition the five-rocket combination might not be powerful enough to reach it.

Secondly, the prospect of launching perhaps hundreds of complex rockets in a few weeks and doing this for perhaps half a dozen launch windows would seem to be highly impractical. The book skims over the technical details and appears to assume that the exercise would be relatively cheap, but this is impossible to credit. Finally, over and above all these objections there is one over-riding problem. The rockets designed to ameliorate the climatic conditions on Mars would have a devastating and perhaps irreversible effect on the Earth's atmosphere.

According to the Allaby–Lovelock scenario, the secondary reason for launching the missiles towards Mars was to get rid of the dangerous solid propellant, which could not be stored indefinitely without the prospect of deterioration and escape into the environment. So the solution is to burn it—in the atmosphere!

Large rockets may be small by comparison with the Earth itself, but it appears that they can have a seeding effect with possible atmospheric results on a very wide scale.

Solid rockets employ as a propellant a rubbery elastomer which contains such substances as ammonium perchlorate, aluminium, a bonding polymer and a curing agent. To spread the combustion products from this propellant across a wide swathe of the southern hemisphere from the proposed launch site in southern India would have unpredictable and almost certainly dangerous results. The chlorine and aluminium would be injected directly into all layers of the atmosphere and might easily change the climate of the whole world for the worse.

It is a pity that the Allaby–Lovelock argument is marred by these technological flaws, for the concept of the greening of Mars is an exciting one. Professor Lovelock is internationally known for his 'Gaia' concept. This puts forward the view that life, far from fitting in with the conditions that it finds on a planet, alters those conditions by a sort of positive feedback to suit its own requirements.

Many scientists see some merit in his arguments, for experience shows that organisms do tend to ameliorate the conditions under which they live. But a more conventional view is that it is the process of natural selection mindlessly producing species which survive best by changing their environmental conditions that is working, rather than the Lovelock's 'superorganism'.

Lovelock postulates that the human race can deliberately use the ability of life forms to adapt their environment to produce a habitable planetary surface and atmosphere on Mars. Once the atmosphere had been warmed to what might be called room temperature by the injection of the CFCs, a series of events could be manipulated to make Mars more and more congenial to man.

The first settlers on Mars would live in an enclosed habitat in which there would be a low-pressure atmosphere with a high content of oxygen. The pressure would be low to avoid a catastrophic escape of the oxygen if the integrity of the building was breached.

Allaby and Lovelock, like the Ames team, visualize the use of algae to lower the albedo of the planet, and so contribute further to the warming process. In a complicated series of arguments, they postulate that it would be possible to grow genetically modified terrestrial plants in the lower martian latitudes, as long as the soil organisms which plants need for the concentration of nutrients were imported as well. By the time the first humans arrived—apparently 14 years after the first CFC rockets—the soil would already be acquiring fertility.

The authors visualize a Mars rapidly becoming green as the settlers plant shrubs such as horse-tails over huge areas of the planet to continue the task of ameliorating the climate.

Meanwhile, food plants for the settlers are grown indoors hydroponically—a technique which has been shown to produce very high yields on Earth. The settlers can leave their secure buildings and walk around on the surface of Mars in ordinary clothing, but need a specially devised breathing apparatus. In fact, pressure suits would be needed until the atmospheric pressure had reached some hundreds of millibars.

The book is an entertainment rather than a scientific treatise but Professor Lovelock has long shown an innovative understanding of the planetary implications of biological evolution and it deserves to be considered as a serious contribution to the debate on how Mars could be colonized.

What methods, other than those outlined in the Ames report, are available to make it happen? It would be futile to try such a mammoth task by industrial methods; no factories built by man could possibly churn out enough oxygen from the martian rocks to have any effect on the composition of the atmosphere.

The genetic engineering path is certainly one that there has been progress on. In one interesting experiment which was aborted because of environmentalists' objections, it was proposed to delete from the genome of a bacterium called *Pseudomonas syringae* a certain section of genes. In greenhouse tests it was shown that when the altered bacteria were sprayed onto potato plants they were better able to survive cold temperatures.

It appeared that the bacteria, which are normally found on potato plants, act as nuclei for ice crystals and therefore actively assist the formation of damaging frost on the plants. When the appropriate genes were deleted the bacteria lost this capacity and therefore it was hoped by the researcher, Steven Lindow, of the University of California at Berkeley, to show that the plants could be protected from frost by this simple piece of genetic engineering.

Taken to the extreme this method would appear to have some relevance to the problems of Mars. It is possible to visualize at some time in the future the production of plants, perhaps with the aid of modified micro-organisms, which would relish the conditions of temperature and aridity to be found on Mars. Desert plants and algae from Earth, modified so that they were completely new species, would then be introduced to Mars on a large scale.

The Greening of Mars envisages something of this sort, with seeds wrapped in organic fertilizer being sprayed over the martian landscape. The plants at first would have to be given genes that would enable them to survive the bitterly cold martian nights and certainly for some generations they would be planted only round the equatorial belt, where temperatures in the daytime can reach 17 °C, although they may be 90° lower than this just before dawn.

In the Ross Desert in Antarctica E Imre Friedmann of Florida State University discovered that bacteria were living in conditions in which probably no other organisms could survive. Apart from the presence of the normal atmospheric oxygen, the Ross Desert conditions are similar to those on Mars, with year-round temperatures well below freezing and very little water. The bacteria were protected from the harshest conditions by existing in crevices near the surface of porous rocks. In their metabolism they leach iron from sandstone and in doing so leave white patches caused by iron depletion. Friedmann suggests that if similar white patches are found in Mars they will be fossil evidence of the existence of micro-organisms in the past.

But even if no such organisms have ever lived on Mars, it may be possible to adapt them to live on the Red Planet, where they would find no competition and would help to raise the oxygen content and pressure of the atmosphere.

One fact unknown to the Ames team is that there are apparently no organic molecules in the martian soil; this was discovered by Viking. The formation of a humus from the dead remains of the early plants will be an essential feature of the first decades of the transformation of the planet.

There still remains the question of the resources available on Mars to explorers and colonizers.

At the 1987 Congress of the International Astronautical Federation Thomas R Meyer, of the Boulder Center for Science and Policy in Colorado, revealed a formidable list of substances that could be manufactured from the martian atmosphere and soil. They included ammonia and methane in addition to the known gases of the atmosphere, together with water, hydrogen, magnesium, titanium, sulphur and hydrogen peroxide. Products which could then be made from these elements and compounds include insulation, storage containers, cement, glass, propellant, plaster, ceramics and explosives.

A fascinating example of the kind of advanced technology that might one day be brought to bear on the problem of terraforming Mars has been described in a number of publications by K Eric Drexler, a computer scientist at the Artificial Intelligence Laboratory of the Massachusetts Institute of Technology in Boston. He calls it nanotechnology because its artefacts would be on a scale where the nanometre (10^{-9} m) is the standard of reference.

He foresees a discipline that employs molecular-sized gears, bearings and rods and even nanocomputers. The nanocomputer would pack 1000 million bytes of information into a space no bigger than a bacterium. The logic would be carried by a series of interlaced rods positioned in a three-dimensional matrix and although the mechanical action would be slower than electronic action, the nanocomputers would be extremely fast.

They would be used to control molecular-sized machines that could both assemble and disassemble artefacts. The assembly would be done by bonding atoms together in chosen patterns. Drexler gives the example of a seamless rocket being built from a vat of raw materials and under the control of a communications network. The result would be an engine with only 10 per cent of the mass of a conventionally built one.

He said in a book he wrote on the subject, *Engines of Creation* (Anchor Press, 1987): 'For all its excellence this engine is fundamentally quite conventional. It has merely replaced dense metal with carefully tailored structures of light, tightly bonded atoms. The final product contains no nanomachinery.' The nanomachines, or 'assemblers', would be able to construct any object that the communications network was programmed to provide. Similarly, 'disassemblers' would be able to strip any object down atom by atom and store in computers the information on exactly how it was made. Assemblers could then take the information from the computers and re-assemble an exact duplicate.

One proposed use for nanotechnology is the production of rocket propellant on the surface of Mars from the local materials. Making hydrogen peroxide or nitric acid would be a simple task for such an advanced technology.

Perhaps more exciting is the prospect of changing the face of the globe by means of more complex chemical reactions. Assemblers could be programmed to reproduce themselves and then, at a certain stage, after enough have been created, they could all begin the task of changing the martian environment. They might, for instance, be seeded over wide areas of the surface and programmed to produce organic molecules which would make the soil fertile.

Alternatively, they could begin to release oxygen and water vapour bound in the rocks and dust of the regolith, so making the environment more congenial by increasing atmospheric pressure and humidity.

Drexler's ideas may be difficult to accept, but they are well enough formulated to satisfy the editors of one of America's leading journals (1981 *Proc. Natl Acad. Sci.* **78** 5275–8). They are an imaginative leap into the future which may or may not be accurate.

If a scientific base were set up on Mars it would undoubtedly have to consist of pressurized habitats of some sort. Early in the settlement of Mars, habitats would have to be covered by a layer of soil as a radiation shield, and excursions to places of interest would be conducted in space suits or in pressurized rovers which would have to include some form of radiation shelter if journeys were too long for an emergency return to base if a solar flare was experienced.

Life support systems would be of the closed cycle type, with all waste products from the base being re-used. Oxygen in the pressurized habitats would be topped up both from local sources and from the greenhouses where food plants would be grown. At least some of the base personnel would have to be expert gardeners. Legumes and cereals would not only provide proteins, carbohydrates, fats, minerals, vitamins and trace elements, but they would also assimilate carbon dioxide and excrete oxygen.

Some of the suggestions for terraforming Mars that have been made appear bizarre, to say the least. That does not mean to say that they are quite out of the question, but one which does appear to demand virtually unbelievable advances in technology while being theoretically possible was put forward by Dr A W G Kunze, of the University of Akron. Kunze's idea is to take an asteroid from the belt which these fragments of stone or ice occupy between Mars and Jupiter and crash it into the surface of Mars. The rationale behind this idea is that if a crater is formed the atmosphere which rushes in to fill it will be much denser than that at ground level, simply by virtue of its being much deeper.

An asteroid with a diameter of 67 km and a density of $3\,\mathrm{g\,cm^{-3}}$, nudged from its heliocentric orbit and crashing into Mars at a velocity of $5.1\,\mathrm{km\,s^{-1}}$, would excavate a crater 41 km deep.

At this depth below the mean martian surface the atmosphere would have a pressure of 500 mb, or half the atmospheric pressure at sea level

The astronaut in the corner of this picture of the Mars base appears to have discovered the fossil of a trilobite. An unmanned aircraft flies overhead and a spacecraft takes off on what is evidently a busy day for the astronauts (NASA).

on Earth. Liquid water would be able to exist as long as the temperature was above 0 °C, and humans would be able to live with normal protective clothing and an oxygen mask. Conditions, in fact, would not be significantly inferior to those on top of high terrestrial mountains.

There might also be plenty of water in the depths of the crater. If the asteroid were made of ice, as some are believed to be, that would in itself dump hundreds of millions of tons of water in the crater, forming a permanent lake. But if the target area were one of those on Mars with subsurface water, there would also be copious springs, streams and even rivers. According to this theory, plants would be able to grow in the bottom of this enormous crater without the benefit of a greenhouse and eventually the atmosphere would come to contain oxygen formed biogenically, as Earth's oxygen was.

Perhaps more likely is the suggestion by various NASA personnel and also some Russian spokesmen that mirrors in space could be used

A manned rover raises dust from the martian surface while astronauts use rocket-propelled manoeuvring units in this imaginative painting of the first martian base. In the background a spacecraft lifts off on the first stage of a journey back to Earth (NASA).

to raise the temperature of the polar regions and in doing so lift the atmospheric pressure by releasing carbon dioxide and water vapour.

The proposal that gigantic mirrors in orbit should be used to light and heat areas of the northern hemisphere of the Earth came from Russian experts back in the 1960s. With much of their agricultural land in northern latitudes with a short growing season the Russians are obviously in the best position to benefit from such an scheme, but Canada and the Scandinavian countries could also find it useful.

The technology for such mirrors has not yet been developed but it is not inconceivable that they will be designed and built for use on Earth before they are employed in martian orbit. They would have to be very large—several square kilometres at least—and this means that the construction will have to be extremely lightweight. A thin plastic sheet with a monomolecular metal coating has been suggested for a deployable type of mirror which could be placed into orbit in a folded form and expanded by means of lightweight struts to the necessary area.

As the mirrors would be in a weightless condition they would not need to be strong enough to support their own weight. They would be something like gossamer and indeed the latest research on a genetically engineered form of spider's web suggests that this natural material might be a candidate for this and other space-related projects.

When an extremely light object with a large area is positioned in space it is subjected to pressure from the light from the Sun. Photons striking the surface impart their energy to it and although each makes only an infinitesimal difference to the object's position, taken together over a protracted period of time billions of impacts can shift any mass. This light pressure, which has a specific impulse measured in millions of seconds, it has been suggested, will one day be a propulsive method for moving large payloads from one planet to another.

The pressure on a huge 'solar sail' attached to the payload will slowly but surely cause it to spiral out from the Earth, for instance, until it reaches the orbit of Mars. In the same way, the space mirrors may be deployed in Earth orbit and then use the pressure of the Sun's light to move them to Mars.

An interesting theoretical contribution to the debate on terraforming Mars was made by Christopher P McKay, of the Department of Astro-Geophysics, University of Colorado, Boulder (1982 *J. Br. Interplanet. Soc.* **35** 427–33).

He pointed out that terraforming Mars would involve two stages— firstly warming the surface and so increasing the surface pressure, and secondly chemically altering the composition of the atmosphere. He went on: 'Estimates for the timescales of the first and second stages are 10^2 and 10^5 years respectively. Constant technological input would be required only during the first stage—that is, the project would require only 100 years of effort, but would not be completed for 100 000 years.'

Since humans are accustomed to measure their personal activities in lifetimes at the most, and governments rarely plan more than a year ahead, these figures may seem daunting. Yet some of mankind's great advances have been made over extended periods. The colonization of the Americas, for instance, took about 300 years. In that case the movement was almost unconscious and was perpetuated by innumerable decisions by governments and individuals of many generations. If a terraforming operation were undertaken there would have to be one conscious decision to commit large resources to the project over decades or even centuries. Large construction efforts, such as the Channel Tunnel or a multi-reactor power station, now occupy as much as a decade or even longer, and the storage of radioactive waste from power generation will have to be monitored for hundreds of years by our descendants. Is it fanciful to imagine that mankind will in future make decisions which will result in actions over periods of equal length?

McKay pointed out that the major uncertainty in the terraforming model was the amount of subsurface volatiles such as nitrogen, oxygen and carbon dioxide. In particular, there was insufficient evidence to determine whether the apparent global shortage of nitrogen and water was real.

He went on: 'The object of terraforming is to alter the environment of another planet so as to improve the chances of survival of an indigenous biology or to allow habitation by most, if not all, terrestrial lifeforms. The resulting system should be stable over long timescales and should require no, or a minimum of, continued technological intervention.'

He preferred 'passive terraforming', in which there was time to observe and learn from the complex interaction of physical and biological factors, to 'technological feats on an unprecedented scale such as alteration of the orbital and physical parameters of a planet or the transport of planetary quantities of materials.' Such proposals, he said, were genuinely beyond the realm of present engineering capabilities. More modest and hence long-term approaches were not only more feasible but more desirable as well.

Relating surface gravity to distance from the Sun, and imposing restraints of 0.5 to 2 atmospheres surface pressure and mean surface temperatures of 260 to 300 K, McKay noted that Mars lies just within or close to the acceptable region. He continued:

> Mars would probably be habitable if it had an Earth-like atmosphere. The decreased surface gravity of Mars results in a more massive atmosphere and an increased greenhouse effect, compensating for its increased distance from the Sun.
>
> It is interesting to note that if the Earth were the same distance from the Sun as Mars, Earth would clearly not be habitable and vice versa. Mars is the only planet which even comes close to satisfying the criteria of biocompatibility.

Discussing the long-term stability of a terraformed Mars, McKay declared:

> Even if a planet can sustain a habitable biosphere by virtue of its mass, distance from the Sun and volatile inventory, the further question of the stability of the system to both external and internal perturbations must be considered. This is particularly important in the case of Mars, owing to the large variations in obliquity and eccentricity which it undergoes.

It is believed that variations in the orbital parameters of Earth are a significant contribution to climatic variation and may well be correlated to the occurrence of ice ages. This is the Milankovich theory, named for the Yugoslavian scientist who first proposed it.

Variations on Mars are much more extreme. The variation in the obliquity leads to a doubling or halving of solar insolation in the climatically sensitive polar regions over a timescale of 100 000 years. Even on Earth a 1 per cent alteration in insolation would result in severe climatic repercussions, resulting in runaway glaciation on the one hand and complete polar melting on the other.

McKay came down firmly on the side of Lovelock's Gaia hypothesis in discussing the 'fairly consistent' long-term stability of the Earth's climate over 3000 million years. He concluded:

> The basic tenet of this terraforming model is that, while alterations in the habitable state of a planet may be accomplished by suitable applications of technology, the ultimate stability and control of the planetary environment is dependent upon the planetary biology.
>
> Furthermore, biology, once established on a planetary scale, acts in such a manner as to maintain the planetary environment at conditions that are beneficial to itself and allow for a complex and diverse biota.

McKay listed the properties of the martian environment which seem most unfavourable to the origin, evolution and sustenance of life, as follows:

1. the general scarcity of water;
2. the low temperatures and extreme temperature variations that occur both diurnally and seasonally;
3. penetration to the surface of ultraviolet radiation between 190 and 300 nanometres wavelength;
4. the presence of strong oxidants in the soil;
5. the low surface pressure and, as a direct result, the exclusion of liquid water as an equilibrium state;
6. the low concentration of N_2 in the atmosphere and the apparent absence of nitrogen in any form in the top regolith.

He stated, however, that all the unfavourable aspects of the martian environment have sufficient variations that any one of them can be eliminated by suitable site selection. The shortage of nitrogen on the martian surface could pose the most severe problem to pioneer martian organisms, since any biotic activity would very likely have to rely on fixation of atmospheric nitrogen at partial pressures of 0.2 millibars.

This is far below the 780 mb of N_2 available at sea level on Earth. However, recent results have indicated that the nitrogen-fixing bacterium, *Beijerinckia lacticogenes*, is capable of using Mars-like levels of atmospheric nitrogen as its sole source of the element.

As far as water is concerned, measurements of the water vapour concentrations in the atmosphere over more than one martian year support a model in which there is a permanent reservoir of water ice at all latitudes poleward of 46 °N and 35 °S. The top surface of the ice reaches the surface at the polar caps and elsewhere is at a depth of between 10 cm and 1 m.

Apart from the implications for man, McKay suggested that research needed to be done on methods for triggering the runaway greenhouse, the volatile abundances on Mars, simple ecosystems which may be the pioneer martian organisms, and the 'diverse and complex planetary ecosystem on which the stability of the final Mars climate may depend'.

To sum up, manned spaceflight has now proved its worth, despite opposition from some academic quarters. Present day critics of an expanding exploration of Mars may prove to be just as wrong in a generation or two.

It is not a matter of blind faith to assert that there will be human exploration and even colonization of the planet Mars. Science and technology are both moving inexorably towards a level at which such a movement will seem both desirable and inevitable. Political and financial decisions may be taken slowly or quickly but in the long run they will not affect the result.

Bibliography

In addition to the publications mentioned below reference has been made frequently to *Spaceflight*, published by the British Interplanetary Society, and to press releases from the National Aeronautics and Space Administration and the Novosti press agency.

Chapter 1

Armitage A 1961 *John Kepler* (London: Nielson)
Gangale T 1988 Lost calendars of Mars *Spaceflight* **30** July 278–83
Koestler A 1959 *The Sleepwalkers* (London: Hutchinson)

Chapter 2

NASA 1974 *Mars as Viewed by Mariner 9. NASA SP-329*
Sheldon C (ed) 1971 *Soviet Space Programs 1966–70* (Washington, DC: US Govt Printing Office)
—— 1976 *Soviet Space Programs 1971–75* (Washington, DC: US Govt Printing Office)

Chapter 3

Baker V R 1982 *The Channels of Mars* (Austin, TX: University of Texas Press)
Ezell E C and Ezell L N 1984 *On Mars. Exploration of the Red Planet 1958–1978. NASA SP-4212*
Jones K L, Arvidson R E, Guiness E A, Bragg S L, Wall S D, Carlston C E and Pidek D G 1979 One Mars year: Viking lander imaging observations *Science* **204** 799–805
Morrison D 1988 The exploration of the Solar System *J. Br. Interplanet. Soc.* **41** 41–7
NASA 1976 *Viking 1, Early Results. NASA SP-408*
—— 1980 *Viking Orbiter Views of Mars. NASA SP-441*
US Geological Survey 1987 Physical properties of the surface materials at the Viking landing site on Mars *US Geological Survey Professional Paper 1389*

Chapter 4

Breus T 1988 *Space Project 'Phobos'. Novosti press release SR00688*
McKenna-Lawlor S M P 1988 Mission to Mars and its moons *New Scientist* **117** 3 March 53–7
NASA 1983 *Planetary Exploration Through Year 2000: A Core Program* Part 1 of a report by the Solar System Exploration Committee of the NASA Advisory Council
—— 1986 *Planetary Exploration Through Year 2000: An Augmented Program* Part 2 of a report by the Solar System Exploration Committee of the NASA Advisory Council
—— 1987a *Mars Observer. Jet Propulsion Laboratory Fact Sheet*
—— 1987b *Mars Sample Return Mission. JPL Fact Sheet*
—— 1988 *NASA Deep Space Network to support Soviet Phobos Mission. NASA press release 88–87*

Chapter 5

Duke M B 1987 Surface activities on the first piloted Mars missions *Paper presented at the 38th IAF Congress* IAF-87-436
French J R 1987 Aerobraking and aerocapture for Mars missions *Paper presented at the 38th IAF Congress*
Menees G P 1987 Aeroassisted-vehicle design studies for a manned Mars mission *Paper presented at the 38th IAF Congress* IAF/IAA-87-433
Page M 1986 Earth-to-Orbit launch vehicles for manned Mars mission application *Manned Mars Missions Working Group Papers* Vol 1, NASA M002
Pritchard E B and Murray R N 1987 Manned Mars mission accommodation by the evolutionary space station *Paper presented at the 38th IAF Congress* IAF-87-438
Ride S K 1987 *Leadership and America's Future in Space* NASA
Roy A E 1988 *Orbital Motion* (Bristol: Adam Hilger) 3rd edn

Chapter 6

Coomes E P, Cuta J M and Webb B J 1986 Pegasus: A multi-megawatt nuclear electric propulsion system *Manned Mars Missions Working Group Papers* Vol 2, NASA M002
Davis H P 1986 A manned Mars mission concept with artificial gravity *Manned Mars Missions Working Group Papers* Vol 1, NASA M002
Duke M B 1986 Report Overview *Manned Mars Missions Working Group Summary Report* NASA M001
French J R 1986 The 'Case for Mars' concept *Manned Mars Missions Working Group Papers* Vol 1, NASA M002
Howe S D and Hynes M V 1986 Anti-matter propulsion: status and prospects *Manned Mars Missions Working Group Papers* Vol 2, NASA M002

Roberts B B 1986 Concept for a manned Mars flyby *Manned Mars Missions Working Group Papers* Vol 1, NASA M002

Tucker M, Meredith O and Brothers B 1986 Space vehicle concepts in manned Mars missions *Manned Mars Missions Working Group Papers* Vol 1, NASA M002

Young A C, Meredith O and Brothers B 1986 Manned Mars flyby mission and configuration concept *Manned Mars Missions Working Group Papers* Vol 1, NASA M002

Chapter 7

Coleman M E 1986 Toxicological safeguards in the manned Mars mission *Manned Mars Missions Working Group Papers* Vol 2, NASA M002

Hamaker J and Smith K 1986 Manned Mars mission cost estimate *Manned Mars Missions Working Group Papers* Vol 2, NASA M002

Johnson P C 1986 Adaptation and readaptation medical concerns of a Mars trip *Manned Mars Missions Working Group Papers* Vol 2, NASA M002

Mandell H C 1987 The cost of landing men on Mars *Paper presented at the 38th IAF Congress*

Nachtwey D S 1986 Manned Mars mission radiation environment and radiobiology *Manned Mars Missions Working Group Papers* Vol 2, NASA M002

Ruppe H O 1987 Expedition to Mars—a baseline mission now *Paper presented at the 38th IAF Congress* IAF-84-198

Santy P A 1986 Manned Mars mission crew factors *Manned Mars Missions Working Group Papers* Vol 2, NASA M002

Schmitt H H 1986 Intermars: user-controlled international management system for missions to Mars *Manned Mars Missions Working Group Papers* Vol 2, NASA M002

US Govt Printing Office 1988 *Soviet Space Programs 1981–87*

Young A C 1986 Mars mission concept and opportunities *Manned Mars Missions Working Group Papers* Vol 1, NASA M002

Chapter 8

Hamaker J and Smith K 1986 Manned Mars mission cost estimate *Manned Mars Missions Working Group Papers* Vol 2, NASA M002

Mandell H C 1981 The cost of landing man on Mars *American Astronautical Society paper AAS 81–251*

Ride S K 1987 *Leadership and America's Future in Space* NASA

Ruppe H O 1984 Expedition to Mars—a baseline mission now *Paper presented at the 35th IAF Congress* IAF-84-198

Schmitt H H 1986 Intermars: user-controlled international management system for missions to Mars *Manned Mars Missions Working Group Papers* Vol 2, NASA M002

Chapter 9

Averner M M and MacElroy R D (ed) 1976 *On the habitability of Mars, an approach to planetary ecosynthesis, NASA SP-414*

Drexler K E 1987 *Engines of Creation* (New York: Anchor)

Lovelock J and Allaby M 1984 *The Greening of Mars* (London: Andrew Deutsch)

Meyer T R 1987 Life science technology for manned Mars missions *Papers presented at the 38th IAF Congress* IAF-87-437

McKay C P 1982 Terraforming Mars *J. Br. Interplanet. Soc.* **35** 427–33

Index

Academy of Sciences (Soviet), 45, 52
Acetaldehyde, 110
Acetone, 110
Acid bottle, 59
Advective heating, 125
Aeroassist Flight Experiment, 76
Aerobraking, 54, 58, 80, 86, 92–4, 108
Aerocapture, 58, 60, 68, 74–6
Aeronomy Orbiter, 55
Aerosol cans, 127
Agena rocket, 15
Air pressure, 125
Albedo, 48, 125, 128
Aleksandrov, Aleksandr, 104
Algae, 124
Allaby, Michael, 125–8
Aluminium, x, 42, 47, 128
Ames Research Center, 75–6, 124, 126, 128–9
Ammonia, 129
Ammonium perchlorate, 128
Anaerobes, 124
Anaesthetic, 105
Antarctica, 110, 129
Antarctic dry valleys, 125
Antimatter, 98
Antiprotons, 98
Aphelion, 6, 65

Apoapsis, 22, 75
Apollo, 20, 65, 68, 70, 75, 77, 82, 93–4, 103, 107, 110, 113, 117–20
8, x
17, 113, 121
Ares, 1
Argillaceous minerals, 43
Argyre, 45
Aristotle, 2
Arizona, 22, 24
Artificial gravity, 82, 93–4, 117
Artificial intelligence, 53, 73
Arrhythmias, 103
Arsia Mons, 22
Ascraeus Mons, 22
Ash falls, 24
Asteroid, 49, 55, 114, 131–2
Asteroidal impacts, 56
Astronomia Nova, 3
Atlas–Agena rocket, 14–15
Atlas–Centaur rocket, 14, 19, 21
Atlas rocket, 15
Atmospheric temperature, 55
Automation, 73

Babakin Design Bureau, 51
Babylon, 1
Bacteria, 124–5
Balebanov, Vyacheslav, 83

Basalt, 31
Balloon, 51–3
Battelle Northwest Laboratory, 96
Beijerinckia lacticogenes, 137
Blue–green algae, 121, 125–6
Bone marrow, 108
Bones, 107
Borman, Frank, 10
Boulder Center for Science and
    Policy, 129
Brahe, Tycho, 2
Brothers, Bobby, 90, 95
Bulgaria, 50
Burroughs, Edgar Rice, 6

Calcium, 42, 107
Canals, 5, 28
Cape Canaveral, 21, 34, 89
Carbohydrates, 131
Carbon, 39–40, 43
    black, 121
    dioxide, x, 12, 20, 27, 40–1, 43,
        45, 54–6, 113, 121, 125,
        127, 131, 133, 135
    monoxide, x, 20, 40, 68, 110
carbon-14, 40
Carbonaceous chondrite
    meteorite, 49
Carr, Michael H, 61
Case for Mars, 78, 94, 111
Cassini, Giovanni Domenico, 4
Celestial mechanics, 22
Centaur, 58, 89
Cereals, 131
Challenger, ix, 55, 88, 117
Channels on Mars, 28, 30, 48
Chaotic terrain, 31, 33
Chemical rockets, 64
Chemotrophy, 40
'Chicken soup', 41
Chlorine, 42, 127–8

Chlorofluorocarbons, 127
Chryse Planitia, 31, 35–6, 43
Climate changes, 56
Climatology Orbiter, 54
Clouds, 45–7
Coleman, Martin E, 110
Comparative planetology, 48, 57
Conjunction missions, 65–6, 70,
    79, 91, 92, 94, 106, 108
Copernicus, 2
Coprates Chasma, 31
Corona, 48, 50
Cosmic rays, 48, 109
Cosmos, 11
Craters, 17, 20–1, 24, 27, 31, 33,
    38, 48, 131
Czechoslovakia, 50

Davis, Hubert P, 93
Deep Space Habitat, 94
Deep Space Network, 45–6
Deimos, ix, 28, 48, 93, 99
Delta *V*, 61, 65, 76, 99
Density of Phobos, 49
Department of Defense, 88
Deuterium/hydrogen ratio, 48
Differentiation, 42, 48
Disarmament agreements, 118
Discovery, 55
DMSP, 55
Drexler, K Eric, 130–1
Dry ice, 45
Duke, Michael B, 78
Dunes, 56
Dust, 48, 55–6, 125, 131
    storms, 6, 12, 22, 46, 126

Earth Return Capsule, 60
Earth Return Vehicle, 60
East Germany, 50
Eccentricity, 135

Ecosphere, 123
Ecosynthesis, 124
Electric propulsion, 61
Electrons, 109
Elysium Mons, 46
EMPIRE, 116
Energiya, ix, 68, 73, 83–4, 89, 96, 114
Engines of Creation, 130
Epicycles, 2
Escape stage, 64
Escape velocity, 30, 104
European Space Agency, 50, 54
Exhaust velocity, 98
Extra-vehicular activity, 103

Fast reactor, 97
Fats, 131
Fertilizer, 129
Finland, 50
Fletcher, James, 121
Florida State University, 129
Fly-by, 95
Fontana, Francesco, 4
French, James, 76, 94
Freon, 110, 113
Friedmann, E Imre, 129
Frost, 129
Fuel cells, 69
Fungi, 124

Gaia, 127–8, 136
Galileo, 3
Gamma radiation, 47
Gamma ray spectrometer, 54–5
Gangale, Thomas, 7
Gas chromatograph/mass spectrometer, 39
Gas chromatography, 40
Gas exchange experiment, 40
Geiger counter, 59

Gemini, 107
Genetic engineering, 125–6, 129
Genome, 129
Geoscience Orbiter, 54
Geosynchronous orbit, 76
Glaciation, 56, 136
Goddard, 82
Goldstone, 45
Grand Canyon, 24
Gravitational assist, 96
Gravitational field, 55
Gravity, 4, 103, 108
Greenhouse effect, 121, 125, 127, 137
Greenhouses, 131–2
Greening of Mars, The, 126, 129

Halley's Comet, 45
Hamaker, Joseph, 118
Hawaii, 22
Health care, 106
Heartbeat, 103
Heatshield, 74, 91, 93
Heavy Lift Launch Vehicle, 68, 70, 73, 85–6, 89, 114–15
Heliocentric orbit, 12, 64–5, 131
Helium, 40, 51
Hellas, 45
Herschel, Sir William, 4
Hess, Seymour L, 43
Horse-tails, 129
Howe, Steven D, 98
Humus, 129
Huygens, C, 4
Hydrated minerals, 33
Hydrazine, 51
Hydrocarbons, 83, 89
Hydrogen, 12, 20, 30, 39, 68, 83, 98, 129
  peroxide, x, 129–30
Hydroponics, 129

Hynes, Michael V, 98
Hypergolic fuels, 21

IAF, 129
ICBMS, 127
Ice, 20, 30, 132
Ice crystals, 129
Infectious diseases, 108
Infrared radiometer, 20, 22
Infrared spectrometer, 20, 22
Infrared thermal mapper, 45
INTELSAT, 113
INTERMARS, 113
Institute of Space Studies, 83
Interplanetary dust, 17
Interplanetary Vehicle System, 58
Ion Beam, 49
Ionosphere, 47
Iran–Iraq war, 120
Irish Republic, 50
Iron, x, 42–3, 47–8, 129
    hydroxides, 43
    oxide, 42
Isopropyl alcohol, 110
Ius Chasma, 30

Japan, 54
Jet Propulsion Laboratory, 19–20,
    22, 35, 37, 45–6, 55, 76, 94,
    96
Johnson, Philip C, 105, 108
Johnson Space Center, 76, 78, 82,
    94, 105, 110–11, 116–17
Jupiter, 1, 3, 33, 131

Kepler, Johannes, 2, 4
Krypton, 40
Kunze, A W G, 131
Kvant, 96

Labelled release experiment, 41

Lake Missoula, 33
Landslides, 30–1
Langley Research Center, 73, 83
Laser, 49
Launch windows, 10–12, 17, 21,
    51, 61, 127
Lava, 24, 31
Laveikin, Aleksandr, 103, 104
Layered terrain, 31, 46
Lee waves, 27
Legumes, 131
LEO, 67, 70, 83–5, 88, 92, 94–7,
    99, 104, 109
Levchenko, Anatoli, 105–6
Lewis Research Center, 96
Lichens, 125–6
Life support systems, 131
Lift/drag ratio, 76
Lima, 49
Lindow, Steven, 129
Linear features, 48
Liquid hydrogen, 92, 119
Liquid water, 123, 132
Los Alamos, 86, 98
Lovelock, James, 126–9, 136
Lowell, Percival, 4–5
LOX, 89, 92
Lunar base, 82
Lunar uplands, 49

Maghemite, 42
Magnesium, 42, 47, 129
Magnetic field, 48, 55
Magnetometer, 54–5
Magnetosphere, 48
Malignant disease, 105, 109
Manarov, Musa, 105–6
Mandell, Humboldt C, 116–18
*Manned Mars Exploration
    Requirements and
    Considerations*, 117

Manned Mars Missions, 86, 94, 118
Mapping reflectance spectrometer, 54
Mariner
    project, 24, 29
    3, 14–15
    4, 14–15, 17, 19–20, 45, 65
    6, 19–21
    7, 10, 19–21
    8, 21
    9, 5, 21–2, 24–8, 42, 46–8, 65
Mars
    1, 9–10, 15–16
    2, 11, 21–2
    3, 11–12, 21
    4, 12
    5, 12–13
    6, 12–13
    7, 12–13
Mars, atmosphere, 10, 12, 17, 20, 25, 27, 29, 35, 43, 47, 54, 56, 68, 75, 129
    calendar for, 7
    climatology, 21, 56, 121
    clouds, 21, 27
    day length on, 7
    fly-by mission, 94
    magnetic field, 17
    mass of, 6
    orbit of, 7, 68, 70, 75, 94
    perihelion, 10
    polar axis of, 6
    polar radius of, 7
    radiation belt, 17
    radius of, 6
    rocks of, 12, 129
    rotation of, 7
    seasons of, 6
    sky, 37
    soil, 40–2, 54, 68, 129
    surface, 22, 54–5, 57, 59, 65, 95, 99, 124
    temperatures, 12, 20, 28, 43, 127
    wind speeds, 43
    year length of, 7, 54
Mars Entry Capsule, 58
Mars Excursion Module, 68, 70, 91
Mars Manoeuvring Vehicles, 93
Mars Observer, 55
Mars Orbiting Vehicle, 58
Mars Rendezvous Vehicle, 59
Mars Underground, 78
Marshall Space Flight Center, 73, 82, 85, 86, 90, 95, 104, 118
Martians, 6
Mass spectrometer, 49
McKay, Christopher P, 134–5
Melas Chasma, 31
MEM, 93
Menees, Gene P, 75–6
Mercury, 1, 7, 107
Meredith, Oliver, 90, 95
Methane, x
Meteorites, 39, 48, 50
Meteorological satellites, 55
Meteorology, 43
Methane, 68, 110, 129
Methylethyl ketone, 110
Meyer, Thomas R, 129
Microgravity, 106
Microorganisms, 125, 129
Milankovich crater, 47
Milankovich theory, 136
Mineralogical map, 55
Minerals, 131
Mir, ix, 96, 103–6
Mirrors in space, 132–4
Mission Control Centre (Soviet), 103

Mission Module, 93
Massachusetts Institute of
    Technology, 130
Molecular oxygen, 47
Monomethyl hydrazine, 21
Moon, x, 1–2, 11, 17, 20, 22, 25,
    55, 74, 81–2, 84, 103, 120,
    121
Murray, Robert N, 73

Nanocomputer, 130
Nanometre, 130
Nanotechnology, 130
NASA, 51, 55–6, 61–2, 69, 73, 76,
    79–80, 82, 86, 88, 93–4, 99,
    103–4, 114–18, 132
    objectives for Mars exploration,
        57
    Solar System Exploration
        Committee, 54–7, 59–60
National Space Policy, 79
Natural selection, 128
Neptune, 1
Nergal, 1
NERVA, 62, 98
Newton, Sir Isaac, 4
Nickel, 43, 48
Nitric acid, 130
Nitrogen, 20, 39–40, 43, 124, 135,
    137
    tetroxide, 21
Noctis Labyrinthus, 30
Nuclear power, 81
Nuclear propulsion, 61
Nuclear thermal rocket, 98
Nutrients, 124, 129

Obliquity, 135–6
Olympus Mons, 22, 23, 46
OMV, 118–19

On the Habitability of Mars, 124,
    126
Opposition, 1, 10, 46, 127
    missions, 65, 67, 70, 79, 83,
        90–2, 106
Optical communications, 81
Orbital transfer vehicle, 60, 76,
    80, 82, 90, 93, 95, 96, 118
Orbital variations, 56
Organic molecules, 39, 129, 131
Osteoarthritis, 108
Oxides of nitrogen, x
Oxygen, x, 12, 20, 28, 30, 39–40,
    68, 81, 83, 119, 124–9, 131–2,
    135
    mask, 132
Ozone, 47, 124, 127

Page, Milton, 85, 88–9
Pathfinder, 79, 81
Pavonis Mons, 22
Payload, 19
Pegasus, 97
Penetrators, 56
Perestroika, 113
Periapsis, 22, 75
Permafrost, 30, 47
Peroxides, 41
Phobos, ix, 28, 45–6, 48–51, 93,
    99
    spacecraft, 13, 46, 48–51
Phosphorus, 124
Photometer, 47
Photons, 134
Photosynthesis, 40–1, 124–5
Plane of the ecliptic, 10
Planetary engineering, 121, 126
*Planetary Exploration Through Year
    2000*, 54, 56
Planetary Observer, 51
Planetary weather, 56

Pluto, 1, 7
Plutonium-238, 69
Poland, 50
Polar caps, 4–5, 7, 20, 26, 45, 54, 121, 125
Polar hood, 46
Polaris, 127
Polar sedimentary layering, 56
Poliakov, Valery, 105
Pollack, James, 37
Poseidon, 127
Potassium, 42, 103
Potato, 129
Presbyopia, 108
Pressure modulator infrared radiometer, 55
Pritchard, E Brian, 73
Proteins, 131
Proton rocket, 11–12, 46
Protons, 109
Protozoa, 124
Pseudomonas syringae, 129
Psychological problems, 110–12
Ptolemaic system, 2
Pyrolytic release experiment, 40
Pyruvic acid, 110

Quakes, 43

Radar altimeter, 54–5
Radiation, 106, 108–9
Radioactive isotopes, 31
Radiometer, 47
Ranger, 14
RCA, 55
Reagan, President, 79
Regional flooding, 56
Regolith, 48–9, 131
Retropropulsion, 21, 68, 75–6, 108

Ride, Sally, 61, 65, 67, 70, 73–6, 78–9, 85, 93, 106, 115, 118
Roberts, Barney B, 94
Robotic exploration, 70
Robotics, 73, 81
Rocket propellant, 68, 70, 130
Romanenko, Yuri, 83, 104, 106
Ross Desert, 129
Rovers, 52–3, 80, 131
RTG, 69
Ruppe, Harry, 115

Sagan, Carl, 37, 86
Sagdeyev, Roald Z, 45, 52, 54, 83, 113
Sample Canister Assembly, 60
Sand dunes, 26, 36, 38
Sample return mission, 54, 57, 80
Santy, Patricia A, 111
Satcom K, 55
Saturn, 1
Saturn I, 119
Saturn V, 8, 73–4, 84, 120
S-band occultation, 22
Schiaparelli, Giovanni, 5, 21
Schmitt, Harrison H, 113, 121
SDV, 88–9, 92, 94, 96, 118–19
Seismic data, 57
Seismometer, 43, 50
Sexuality, 111
Shield volcanoes, 56
Silicon, 42
  dioxide, 42
Skatole, 110
Skylab, 84, 106–7
Slayton, Deke, 110
Smith, Keith, 118–19
Snow, 20
Soffen, Gerald A, 34
Sol, 7

Solar energy, 122
  flares, 48, 65, 109, 131
  gamma ray bursts, 48
  luminosity changes, 56
  oscillations, 48
  panels, 46
  power, 117
  sails, 86, 134
  wind, 48–9, 56
Solar Particle Events, 50, 109
Solar System, x, 69, 79, 106, 120, 123
Solid propellant, 127
Soviet space shuttle, 105
Soyuz, 96, 110
  TM2, 103
  TM3, 103
  TM4, 105
Space Adaptation Syndrome, 107
Space Shuttle, 45, 60, 68, 73–4, 76, 84–5, 88–9, 96, 106–7, 116–18
Space station, 60, 70, 73, 76, 83, 90–1, 94–5, 106, 114–19
Space suits, 131
Specific impulse, 61, 73, 81, 83, 97–8, 134
Spider's web, 134
Sprint missions, 70, 73–4, 106, 115
Sputnik, 1, 8
Stafford, Tom, 110
Stickney crater, 49
Stockholm International Peace Research Institute, 120
Storm shelter, 109
Sulphur, 42, 47, 129
Sun, 1–2, 7, 10, 30, 50–1, 111, 123, 135
Superoxides, 41
Sweden, 50

Switzerland, 50
Swing-by manoeuvre, 104
Synchronous rotation, 48
Syrtis Major, 4
Syrtis Major Planitia, 46

Tectonic movement, 24
Terraforming, 126, 130–1, 134–6
Tharsis Ridge, 46
Thermal-emission spectrometer, 55
Thermal infrared radiometer/spectrometer, 54
TIROS, 55
Titan–Centaur rocket, 34
Titanium, 42, 129
Tithonium Chasma, 30
Titov, Vladimir, 105–6
TOS, 55
Toxic gases, 110
Trace elements, 131
Transfer orbit, 9, 65
Transfer Orbit Stage, 55
Trans-Mars trajectory, 46
Trichloroethanetoluene, 110
Trident, 127
Tucker, Michael, 90
TV cameras, 19, 21, 39, 43–4, 50
Tyuratam, 9, 11

Ultraviolet, 30, 39, 124–6
  spectrometer, 20, 22
  spectrometer/photometer, 54
University of Akron, 131
University of California, 129
University of Colorado, 134
Upper atmosphere, 56
Uranus, 1, 4

Valles Marineris, 24, 28, 30–1, 46
Van Allen belts, 109

Vegetation, 26
Venera, 10, 65
Venus, 1–2, 10, 45, 51, 55, 65, 74, 96, 104
Venus swing-by manoeuvre, 65, 67, 96, 104
Very Long Baseline Interferometer, 46, 50
Viking, 29, 34, 42–3, 45, 47–8, 55, 65, 129
  1, 31, 34–8
  2, 10, 37–8
  orbiters, 42, 44
Visual/infrared spectrometer, 55
Vitamins, 131
Volcanoes, 22–5, 28, 41–2
Voyager, 69

War of the Worlds, The, 6
Water, x, 24, 27, 29–31, 33, 42–3, 54–5, 68, 129, 135, 137

Water vapour, 12, 30, 40, 45, 47, 48, 54–5, 121, 125, 131, 133, 137
Wave of darkening, 4, 26
Weightlessness, 67, 103, 105–8
Welles, Orson, 6
Wells, H G, 6
Wind
  deposition, 25
  erosion, 25
  power, 69

X-ray fluorescence spectrometer, 50
X-ray frequencies, 50
Xylene, 110

Yevpatoriya, 46
Young, Archie, 95, 104

Zond
  2, 11
  3, 11